OLIVER SACKS

The Man Who Mistook His Wife for a Hat

Dr. Oliver Sacks, referred to by *The New York Times* as "the poet laureate of medicine," spent more than fifty years working as a neurologist and writing books about the neurological predicaments and conditions of his patients, including *Musicophilia*, *A Leg to Stand On*, and *Awakenings*. Over the years, he received many awards, including honors from the Guggenheim Foundation, the National Science Foundation, the American Academy of Arts and Letters, the American Academy of Arts and Sciences, and the Royal College of Physicians. His memoir, *On the Move*, was published shortly before his death in August 2015. For more information, please visit www.oliversacks.com.

ALSO BY OLIVER SACKS

Migraine

Awakenings

A Leg to Stand On

Seeing Voices

An Anthropologist on Mars

The Island of the Colorblind

Uncle Tungsten

Oaxaca Journal

Musicophilia

The Mind's Eye

Hallucinations

On the Move

Gratitude

The River of Consciousness

Everything in Its Place

The Man Who Mistook His Wife for a Hat

The Man Who Mistook His Wife for a Hat

And Other Clinical Tales

Oliver Sacks

VINTAGE BOOKS

A Division of Penguin Random House LLC

New York

FIRST VINTAGE BOOKS EDITION, SEPTEMBER 2021

The Library of Congress has cataloged the
Summit Books edition as follows:
Names: Sacks, Oliver, 1933–2015.
Title: The man who mistook his wife for a hat
and other clinical tales / Oliver Sacks.
Description: First edition. | New York : Summit Books, c1985.
Identifiers: LCCN 85017220
Subjects: Neurology—Anecdotes.
Classification: LCC RC351 .S195 1985 | DDC 616.8—dc23
LC record available at https://lccn.loc.gov/85017220

Vintage Books Trade Paperback ISBN: 978-0-593-46667-4
eBook ISBN: 978-0-593-46668-1

www.vintagebooks.com

Printed in the United States of America
10 9

To Leonard Shengold, M.D.

To talk of diseases is a sort of *Arabian Nights* entertainment.

—William Osler

The physician is concerned [unlike the naturalist] . . . with a single organism, the human subject, striving to preserve its identity in adverse circumstances.

—Ivy McKenzie

Contents

Preface (2013)

Both of my parents were physicians, and I grew up in a house full of medical stories. At dinner, my mother or father would often tell stories of patients they had seen that day—stories of lives whose course had been cut across by disease or injury. (Sometimes by neurological disease or injury, for both my parents had trained as neurologists before settling on other specialties.) And though, as a schoolboy, I was drawn to chemistry and then to botany and marine biology, it was perhaps inevitable that I would finally gravitate to medicine, with its study and stories of people.

When I became a medical student, as two of my older brothers had, it was the patients I saw, their predicaments and their stories, that gripped my imagination, and these experiences imprinted themselves upon me indelibly. Lectures and textbooks, abstracted from living experience, left almost no impression. I was, however, strongly drawn to the case histories that abounded in the nineteenth-century medical literature—rich, detailed descriptions of patients with neurological or psychiatric problems.

In 1966, as a young physician, I came across the patients I would later describe in *Awakenings*. Their situation was unique in many ways: these people, however different they were as individuals, were all locked into a near-catatonic state by the same disease and had all been incarcerated in a chronic disease hospital for decades. Their "awakenings" from this frozen, Rip van Winkle state, the reclaiming of life made possible by the new drug L-dopa, could not be reduced to surveys or numbers; this required individual, highly personal narratives.

In *The Man Who Mistook His Wife for a Hat*, by contrast, I described, after twenty years of practicing medicine, patients with a wide variety of neurological disorders, some long-standing, some not. A few of these patients, such as Dr. P., the man who mistook his wife for a hat, were able to live fairly full lives outside institutions, and I would visit them in their own homes, their own personal settings.

Nearly thirty years have passed since *Hat* was first published, and some of the patients I described are still alive and thriving. Witty Ticcy Ray, whom I originally saw in 1971, has continued to live a full life despite his Tourette's, and we are frequently in touch. Ray inspired in me an intense, lifelong interest in Tourette's, and I have since written about many other people with the syndrome, including a lengthy case history (originally titled "The World's Only Flying Touretter Surgeon") published as "A Surgeon's Life" in *An Anthropologist on Mars*.

The patients I described in *Hat* are never far from mind, and I continue to make new connections to their stories. Lilian Kallir, a renowned pianist, wrote to me some fifteen years after *Hat* was published, saying that she had lost the ability to recognize objects around her. She compared herself with Dr. P., though her ways of dealing with her problems of visual recognition were quite different from his. Lilian had a condition called posterior cortical atrophy, a term introduced years after *Hat* was published, to describe a specific Alzheimer-like syndrome. So although I could not provide a specific diagnosis for Dr. P. at the time, encountering Lilian years later helped to provide an answer.

Jimmie, "The Lost Mariner," showed me what life was like for someone with a profound amnesia, which I later explored in other patients, like Greg ("The Last Hippie" in *An Anthropologist on Mars*) and Clive Wearing, the conductor whose story I tell in *Musicophilia*. It is only by accumulating case histories of people with similar syndromes, comparing and contrasting them, that one can more fully understand the mechanisms involved and their resonances for an individual life.

"Reminiscence," the story of two old ladies with musical hallucinations, inspired me to undertake a wider survey of such hallucinations (in *Musicophilia*) and then of hallucinations generally (in *Hallucinations*). Phantom limbs, described briefly in *Hat*, are considered at length in *Hallucinations*. And the final chapter in *Hat*, "The Autist Artist," led to much longer case histories of Stephen Wiltshire, another autistic savant, and Temple Grandin, a remarkable woman with Asperger's syndrome (both of which were published in *An Anthropologist on Mars*).

The Man Who Mistook His Wife for a Hat, written in the early 1980s, contains a number of words which are, thankfully, outmoded today. "Idiot savant," "moron," "simpleton," "retardate," and the like, were the terms of the day, and as such I have left them here. Similarly, patients were often regarded then as "psychotic," when today other terms would apply. "Asperger's syndrome" and even "Alzheimer's disease" had not yet entered the medical lexicon.

I no longer agree with some of the things I wrote in *Hat*, and in many cases I have, over time, come to look at these patients in, I hope, a more nuanced way. But for me, they all remain alive, their stories expanding and revising as all of ours do.

The writing of case histories, presenting not only the effects of disease but the whole living reality of being a patient, reached great heights in the nineteenth century, but by the late twentieth century, with the rise of a more technological and quantitative medicine, it had become nearly extinct. Thus when I came to publish my own case histories in the 1970s and 1980s, it was virtually impossible to do so in medical journals, which required charts and tables, and "objective" language. Longer, more personal, detailed case histories were considered archaic and "unscientific." This is beginning to change again—many medical schools have introduced courses in Narrative Medicine, and whole generations of younger neurologists see the case history as a crucial part of medicine. *The Man Who Mistook His Wife for a Hat* is often credited for playing a part in this revival of the tradition of case history, and I like to think that is so.

With the rise of neuroscience and all its wonders, it is even more important now to preserve the personal narrative, to see every patient as a unique being with his own history and strategies for adapting and surviving. Though our knowledge and insights may evolve and change, the phenomenology of human sickness and health remains fairly constant, and case history, careful and detailed descriptions of individual patients, can never become obsolete.

O. W. S.
New York

Preface to the Original Edition (1985)

"The last thing one settles in writing a book," Pascal observes, "is what one should put in first." So, having written, collected and arranged these strange tales, having selected a title and two epigraphs, I must now examine what I have done—and why.

The doubleness of the epigraphs, and the contrast between them—indeed, the contrast which Ivy McKenzie draws between the physician and the naturalist—corresponds to a certain doubleness in me: that I feel myself a naturalist and a physician both; and that I am equally interested in diseases and people; perhaps, too, that I am equally, if inadequately, a theorist and dramatist, am equally drawn to the scientific and the romantic, and continually see both in the human condition, not least in that quintessential human condition of sickness—animals get diseases, but only man falls radically into sickness.

My work, my life, is all with the sick—but the sick and their sickness drive me to thoughts which, perhaps, I might otherwise not have. So much so that I am compelled to ask, with Nietzsche: "As for sickness: are we not almost tempted to ask whether we could get along without it?"—and to see the questions it raises as fundamental in nature. Constantly my patients drive me to question, and constantly my questions drive me to patients—thus in the stories or studies which follow there is a continual movement from one to the other.

Studies, yes; why stories, or cases? Hippocrates introduced the historical conception of disease, the idea that diseases have a course, from their first intimations to their climax or crisis, and thence to their happy or fatal resolution. Hippocrates thus introduced the case history, a description, or depiction, of the natural history of disease—precisely expressed by the old word "pathology." Such histories are a form of natural history—but they tell us nothing about the individual and *his* history; they convey nothing of the person and the experience of the person as he faces and struggles to survive his disease. There is no "subject" in a narrow case history; modern case histories allude to the subject in a cursory phrase ("a trisomic albino female of 21"), which could as well apply to a rat as a human being. To restore the human subject at the center—the suffering, afflicted, fighting human subject—we must deepen a case history to a narrative or tale; only then do we have a "who" as well as a "what," a real person, a patient, in relation to disease—in relation to the physical.

The patient's essential being is very relevant in the higher reaches of neurology, and in psychology; for here the patient's personhood is essentially involved, and the study of disease and of identity cannot be disjoined. Such disorders, and their depiction and study, indeed entail a new discipline, which we may call the "neurology of identity," for it deals with the neural foundations of the self, the age-old problem of mind and brain. It is possible that there must of necessity be a gulf, a gulf of category, between the psychical and the physical; but studies and stories pertaining simultaneously and inseparably to both (and it is these which especially fascinate me, and which, on the whole, I present here) may nonetheless serve to bring them nearer, to bring us to the very intersection of mechanism and life, to the relation of physiological processes to biography.

The tradition of richly human clinical tales reached a high point in the nineteenth century, and then declined, with the advent of an impersonal neurological science. Alexander Luria wrote: "The power to describe, which was so common to the great nineteenth-century neurologists and psychiatrists, is almost gone now. . . . It must be revived." His own late works, such as *The Mind of a Mnemonist* and *The Man with a Shattered World*, are attempts to revive this lost tradition. Thus the case histories in this book hark back to an ancient tradition: to the nineteenth-century tradition of which Luria speaks; to the tradition of the first medical historian, Hippocrates; and to that universal and prehistorical tradition by which patients have always told their stories to doctors.

Classical fables have archetypal figures—heroes, victims, martyrs, warriors. Neurological patients are all of these—and in the strange tales told here they are also something more. How, in these mythical or metaphorical terms, shall we categorize the "lost mariner," or the other strange figures in this book? We may say they are travelers to unimaginable lands— lands of which otherwise we should have no idea or conception. This is why their lives and journeys seem to me to have a quality of the fabulous, why I have used Osler's *Arabian Nights* image as an epigraph, and why I feel compelled to speak of tales and fables as well as cases. The scientific and the romantic in such realms cry out to come together—Luria liked to speak here of "romantic science." They come together at the intersection of fact and fable, the intersection which characterizes (as it did in my book *Awakenings*) the lives of the patients here narrated.

But what facts! What fables! To what shall we compare them? We may not have any existing models, metaphors or myths. Has the time perhaps come for new symbols, new myths?

———

Eight of the chapters in this book have already been published: "The Lost Mariner," "Hands," "The Twins," and "The Autist Artist" in the *New York Review of Books* (1984 and 1985), and "Witty Ticcy Ray," "The Man Who Mistook His Wife for a Hat," and "Reminiscence" in the *London Review of Books* (1981, 1983, 1984)—where the briefer version of the last was called "Musical Ears." "On the Level" was published in *The Sciences* (1985). A very early account of one of my patients—the "original" of Rose R. in *Awakenings* and of Harold Pinter's Deborah in *A Kind of Alaska,* inspired by that book—is to be found in "Incontinent Nostalgia" (originally published as "Incontinent Nostalgia Induced by L-Dopa" in the *Lancet* of Spring 1970). Of my four "Phantoms," the first two were published as "clinical curios" in the *British Medical Journal* (1984). Two short pieces are taken from previous books: "The Man Who Fell Out of Bed" is excerpted from *A Leg to Stand On,* and "The Visions of Hildegard" from *Migraine.* The remaining twelve pieces are unpublished and entirely new, and were all written during the autumn and winter of 1984.

I owe a very special debt to my editors: first to Robert Silvers of the *New York Review of Books* and Mary-Kay Wilmers of the *London Review of Books*; then to Kate Edgar, Jim Silberman of Summit Books in New York, and Colin Haycraft of Duckworth's in London, who between them did so much to shape the final book.

Among my fellow neurologists I must express special gratitude to the late Dr. James Purdon Martin, to whom I showed videotapes of "Christina" and "Mr. MacGregor" and with whom I discussed these patients fully—"The Disembodied Lady" and "On the Level" express this indebtedness; to Dr. Michael Kremer, my former "chief" in London, who in response to *A Leg to Stand On* (1984) described a very similar case of his own—these are bracketed together now in "The Man Who Fell Out of Bed"; to Dr. Donald Macrae, whose extraordinary case of visual agnosia, almost comically similar to my own, was only discovered, by accident, two years after I had written my own piece—it is excerpted in a postscript to "The Man Who Mistook His Wife for a Hat"; and, most especially, to my close friend and colleague, Dr. Isabelle Rapin, in New York, who discussed many cases with me; she introduced me to Christina (the "disembodied lady"), and had known José, the "autist artist," for many years when he was a child.

I wish to acknowledge the selfless help and generosity of the patients (and, in some cases, the relatives of the patients) whose tales I tell here—who, knowing (as they often did) that they themselves might not be able to be helped directly, yet permitted, even encouraged, me to write of their lives, in the hope that others might learn and understand, and, one day, perhaps be able to cure. As in *Awakenings*, names and some circumstantial details have been changed for reasons of personal and professional confidence, but my aim has been to preserve the essential "feeling" of their lives.

Finally, I wish to express my gratitude—more than gratitude—to my own mentor and physician, Leonard Shengold, to whom I dedicate this book.

O. W. S.
New York

PART ONE

Losses

Introduction

Neurology's favorite word is "deficit," denoting an impairment or incapacity of neurological function: loss of speech, loss of language, loss of memory, loss of vision, loss of dexterity, loss of identity and myriad other lacks and losses of specific functions or faculties. For all of these dysfunctions (another favorite term), we have privative words of every sort—Aphonia, Aphemia, Aphasia, Alexia, Apraxia, Agnosia, Amnesia, Ataxia—a word for every specific neural or mental function of which patients, through disease, or injury, or failure to develop, may find themselves partly or wholly deprived.

The scientific study of the relationship between brain and mind began in 1861, when Broca, in France, found that specific difficulties in the expressive use of speech, aphasia, consistently followed damage to a particular portion of the left hemisphere of the brain. This opened the way to a cerebral neurology, which made it possible, over the decades, to "map" the human brain, ascribing specific powers—linguistic, intellectual, perceptual, etc.—to equally specific "centers" in the brain. Toward the end of the century it became evident to more acute observers—above all to Freud, in his book *Aphasia*—that this sort of mapping was too simple, that all mental performances had an intricate internal structure, and must have an equally complex physiological basis. Freud felt this, especially, in regard to certain disorders of recognition and perception,

for which he coined the term "agnosia." All adequate understanding of aphasia or agnosia would, he believed, require a
new, more sophisticated science.

The new science of brain/mind which Freud envisaged
came into being in the Second World War, in Russia, as the
joint creation of A. R. Luria (and his father, R. A. Luria), Leon-
tev, Anokhin, Bernstein and others, and was called by them
"neuropsychology." The development of this immensely fruit-
ful science was the lifework of A. R. Luria, and considering
its revolutionary importance, it was somewhat slow in reach-
ing the West. It was set out, systematically, in a monumen-
tal book, *Higher Cortical Functions in Man* (translated into
English in 1966) and, in a wholly different way, in a biography
or "pathography"—*The Man with a Shattered World* (translated into English in 1972). Although these books were almost
perfect in their way, there was a whole realm which Luria had
not touched. *Higher Cortical Functions in Man* treated only
those functions which appertained to the left hemisphere of
the brain; similarly, Zazetsky, the subject of *The Man with a
Shattered World*, had a huge lesion in the left hemisphere—the
right was intact. Indeed, the entire history of neurology and
neuropsychology can be seen as a history of the investigation
of the left hemisphere.

One important reason for the neglect of the right, or "minor,"
hemisphere, as it has been called, is that while it is easy to
demonstrate the effects of variously located lesions on the left
side, the corresponding syndromes of the right hemisphere are
much less distinct. It was presumed, usually contemptuously,
to be more "primitive" than the left, the latter being seen as
the unique flower of human evolution. And in a sense this is
correct: the left hemisphere is more sophisticated and special-
ized, a very late outgrowth of the primate, and especially the

hominid, brain. On the other hand, it is the right hemisphere which controls the crucial powers of recognizing reality which every living creature must have in order to survive. The left hemisphere, like a computer tacked onto the basic creatural brain, is designed for programs and schematics; and classical neurology was more concerned with schematics than with reality, so that when, at last, some of the right-hemisphere syndromes emerged, they were considered bizarre.

There had been attempts in the past—for example, by Gabriel Anton in the 1890s and Otto Pötzl in 1928—to explore right-hemisphere syndromes, but these attempts themselves had been bizarrely ignored. In *The Working Brain*, one of his last books, Luria devoted a short but tantalizing section to right-hemisphere syndromes, ending:

> These still completely unstudied defects lead us to one of the most fundamental problems—to the role of the right hemisphere in direct consciousness. . . . The study of this highly important field has been so far neglected. . . . It will receive a detailed analysis in a special series of papers . . . in preparation for publication.

Luria did, finally, write some of these papers, in the last months of his life, when mortally ill. He never saw their publication, nor were they published in Russia. He sent them to Richard Gregory in England, and they will appear in Gregory's forthcoming *Oxford Companion to the Mind*.

Inner difficulties and outer difficulties match each other here. It is not only difficult, it is impossible, for patients with certain right-hemisphere syndromes to know their own problems—a peculiar and specific "anosognosia," as Babinski called it. And it is singularly difficult, for even the most sensitive observer, to picture the inner state, the "situation," of such patients, for this is almost unimaginably remote from anything he himself has ever known. Left-hemisphere syndromes, by contrast, are relatively easily imagined. Although right-hemisphere syndromes are as common as left-hemisphere syndromes—why should they not be?—we will find a thousand descriptions of left-hemisphere syndromes in the neurological and neuropsychological literature for every description of a right-hemisphere syndrome. It is as if such syndromes were somehow alien to the whole temper of neurology. And yet, as Luria says, they are of the most fundamental importance. So much so that they may demand a new sort of neurology, a "personalistic," or (as Luria liked to call it) a "romantic," science; for the physical foundations of the *persona*, the self, are here revealed for our study. Luria thought a science of this kind would be best introduced by a story—a detailed case history of a man with a profound right-hemisphere disturbance, a case history which would at once be the complement and opposite of "the man with a shattered world." In one of his last letters to me, he wrote, "Publish such histories, even if they are just sketches. It is a realm of great wonder." I must confess to being especially intrigued by these disorders, for they open realms, or promise realms, scarcely imagined before, pointing to an open and more spacious neurology and psychology, excitingly different from the rather rigid and mechanical neurology of the past.

It is, then, less deficits, in the traditional sense, which have engaged my interest than neurological disorders affecting the self. Such disorders may be of many kinds—and may arise from excesses, no less than impairments, of function, and it seems reasonable to consider these two categories separately. But it must be said from the outset that a disease is never a mere loss or excess; that there is always a reaction on the part of the affected organism or individual, to restore, to replace, to compensate for and to preserve its identity, however strange the means may be—and to study or influence these means, no less than the primary insult to the nervous system, is an essential part of our role as physicians. This was powerfully stated by Ivy McKenzie:

> For what is it that constitutes a "disease entity" or a "new disease"? The physician is concerned not, like the naturalist, with a wide range of different organisms theoretically adapted in an average way to an average environment, but with a single organism, the human subject, striving to preserve its identity in adverse circumstances.

This dynamic, this "striving to preserve identity," however strange the means or effects of such striving, was recognized in psychiatry long ago—and, like so much else, is especially associated with the work of Freud. Thus, the delusions of paranoia were seen by him not as primary but as attempts (however misguided) at restitution, at reconstructing a world reduced by complete chaos. In precisely the same way, McKenzie wrote:

> The pathological physiology of the Parkinsonian syndrome is the study of *an organised chaos*, a chaos induced in the first instance by destruction of important integrations, and reorganised on an unstable basis in the process of rehabilitation.

As *Awakenings* was the study of "an organized chaos" produced by a single if multiform disease, so what now follows is a series of similar studies of the organized chaoses produced by a great variety of diseases.

In this first section, "Losses," the most important case, to my mind, is that of a special form of visual agnosia: "The Man Who Mistook His Wife for a Hat." I believe it to be of fundamental importance. Such cases constitute a radical challenge to one of the most entrenched axioms or assumptions of classical neurology—in particular, the notion that brain damage, any brain damage, reduces or removes the "abstract and categorical attitude" (in Kurt Goldstein's term), reducing the individual to the emotional and concrete. (A very similar thesis was made by Hughlings Jackson in the 1860s.) Here, in the case of Dr. P., we see the very opposite of this—a man who has (albeit only in the sphere of the visual) wholly lost the emotional, the concrete, the personal, the "real" . . . and been reduced, as it were, to the abstract and the categorical, with consequences of a particularly preposterous kind. What would Hughlings Jackson and Goldstein have said of *this*? I have often, in imagination, asked them to examine Dr. P., and then said, "Gentlemen! What do you say *now*?"

1

The Man Who Mistook His Wife for a Hat

D r. P. was a musician of distinction, well known for many years as a singer and then, at the local School of Music, as a teacher. It was here, in relation to his students, that certain strange problems were first observed. Sometimes a student would present himself, and Dr. P. would not recognize him; or, specifically, would not recognize his face. The moment the student spoke, he would be recognized by his voice. Such incidents multiplied, causing embarrassment, perplexity, fear—and, sometimes, comedy. For not only did Dr. P. increasingly fail to see faces, but he saw faces when there were no faces to see: genially, Magoo-like, when in the street he might pat the heads of water hydrants and parking meters, taking these to be the heads of children; he would amiably address carved knobs on the furniture and be astounded when they did not reply. At first these odd mistakes were laughed off as jokes, not least by Dr. P. himself. Had he not always had a quirky sense of humor and been given to Zen-like paradoxes and jests? His musical powers were as dazzling as ever; he did not feel ill—he had never felt better; and the mistakes were so ludicrous—and so ingenious—that they could hardly be serious or betoken anything serious. The notion of there being "something the matter" did not emerge until some three years later, when diabetes developed. Well aware that diabetes could affect his eyes, Dr. P. consulted an ophthalmologist, who took a careful history and examined his eyes closely. "There's nothing the matter with your eyes," the doctor concluded. "But there is trouble with the visual parts of your brain. You don't need my help; you must see a neurologist." And so, as a result of this referral, Dr. P. came to me.

It was obvious within a few seconds of meeting him that there was no trace of dementia in the ordinary sense. He was a man of great cultivation and charm who talked well and fluently, with imagination and humor. I couldn't think why he had been referred to our clinic.

And yet there was something a bit odd. He faced me as he spoke, was oriented towards me, and yet there was something the matter—it was difficult to formulate. He faced me with his *ears*, I came to think, but not with his eyes. These, instead of looking, gazing, at me, "taking me in," in the normal way, made sudden strange fixations—on my nose, on my right ear, down to my chin, up to my right eye—as if noting (even studying) these individual features, but not seeing my whole face, its changing expressions, "me," as a whole. I am not sure that I fully realized this at the time—there was just a teasing strangeness, some failure in the normal interplay of gaze and expression. He saw me, he scanned me, and yet . . .

"What seems to be the matter?" I asked him at length.

"Nothing that I know of," he replied with a smile, "but people seem to think there's something wrong with my eyes."

"But *you* don't recognize any visual problems?"

"No, not directly, but I occasionally make mistakes."

I left the room briefly to talk to his wife. When I came back, Dr. P. was sitting placidly by the window, attentive, listening rather than looking out. "Traffic," he said, "street sounds, distant trains—they make a sort of symphony, do they not? You know Honegger's *Pacific 231*?"

What a lovely man, I thought to myself. How can there be anything seriously the matter? Would he permit me to examine him?

"Yes, of course, Dr. Sacks."

I stilled my disquiet, his perhaps, too, in the soothing routine of a neurological exam—muscle strength, coordination, reflexes, tone. . . . It was while examining his reflexes—a trifle abnormal on the left side—that the first bizarre experience occurred. I had taken off his left shoe and scratched the sole of his foot with a key—a frivolous-seeming but essential test of a reflex—and then, excusing myself to screw my ophthalmoscope together, left him to put on the shoe himself. To my surprise, a minute later, he had not done this.

"Can I help?" I asked.

"Help what? Help whom?"

"Help you put on your shoe."

"Ach," he said, "I had forgotten the shoe," adding, *sotto voce*, "The shoe? The shoe?" He seemed baffled.

"Your shoe," I repeated. "Perhaps you'd put it on."

He continued to look downwards, though not at the shoe, with an intense but misplaced concentration. Finally his gaze settled on his foot: "That is my shoe, yes?"

Did I mis-hear? Did he mis-see?

"My eyes," he explained, and put a hand to his foot. "*This* is my shoe, no?"

"No, it is not. That is your foot. *There* is your shoe."

"Ah! I thought that was my foot."

Was he joking? Was he mad? Was he blind? If this was one of his "strange mistakes," it was the strangest mistake I had ever come across.

I helped him on with his shoe (his foot), to avoid further complication. Dr. P. himself seemed untroubled, indifferent, maybe amused. I resumed my examination. His visual acuity was good: he had no difficulty seeing a pin on the floor, though sometimes he missed it if it was placed to his left.

He saw all right, but what did he see? I opened out a copy of the *National Geographic* magazine and asked him to describe some pictures in it.

His responses here were very curious. His eyes would dart from one thing to another, picking up tiny features, individual features, as they had done with my face. A striking brightness, a color, a shape would arrest his attention and elicit comment—but in no case did he get the scene-as-a-whole. He failed to see the whole, seeing only details, which he spotted like blips on a radar screen. He never entered into relation with the picture as a whole—never faced, so to speak, *its* physiognomy. He had no sense whatever of a landscape or scene.

I showed him the cover, an unbroken expanse of Sahara dunes.

"What do you see here?" I asked.

"I see a river," he said. "And a little guest house with its terrace on the water. People are dining out on the terrace. I see colored parasols here and there." He was looking, if it was "looking," right off the cover into mid-air and confabulating nonexistent features, as if the absence of features in the actual picture had driven him to imagine the river and the terrace and the colored parasols.

I must have looked aghast, but he seemed to think he had done rather well. There was a hint of a smile on his face. He also appeared to have decided that the examination was over and started to look around for his hat. He reached out his hand and took hold of his wife's head, tried to lift it off, to put it on. He had apparently mistaken his wife for a hat! His wife looked as if she was used to such things.

I could make no sense of what had occurred in terms of

conventional neurology (or neuropsychology). In some ways he seemed perfectly preserved, and in others absolutely, incomprehensibly devastated. How could he, on the one hand, mistake his wife for a hat and, on the other, function as, apparently, he still did, as a teacher at the Music School?

I had to think, to see him again—and to see him in his own familiar habitat, at home.

A few days later I called on Dr. P. and his wife at home, with the score of the *Dichterliebe* in my briefcase (I knew he liked Schumann), and a variety of odd objects for the testing of perception. Mrs. P. showed me into a lofty apartment, which recalled fin-de-siècle Berlin. A magnificent old Bösendorfer stood in state in the center of the room, and all around it were music stands, instruments, scores. . . . There were books, there were paintings, but the music was central. Dr. P. came in, a little bowed, and, distracted, advanced with outstretched hand to the grandfather clock, but, hearing my voice, corrected himself, and shook hands with me. We exchanged greetings and chatted a little of current concerts and performances. Diffidently, I asked him if he would sing.

"The *Dichterliebe*!" he exclaimed. "But I can no longer read music. You will play them, yes?"

I said I would try. On that wonderful old piano even my playing sounded right, and Dr. P. was an aged but infinitely mellow Fischer-Dieskau, combining a perfect ear and voice with the most incisive musical intelligence. It was clear that the Music School was not keeping him on out of charity.

Dr. P.'s temporal lobes were obviously intact: he had a wonderful musical cortex. What, I wondered, was going on in his parietal and occipital lobes, especially in those areas where visual processing occurred? I carry the Platonic solids in my neurological kit and decided to start with these.

"What is this?" I asked, drawing out the first one.

"A cube, of course."

"Now this?" I asked, brandishing another.

He asked if he might examine it, which he did swiftly and systematically: "A dodecahedron, of course. And don't bother with the others—I'll get the icosahedron, too."

Abstract shapes clearly presented no problems. What about faces? I took out a pack of cards. All of these he identified instantly, including the jacks, queens, kings, and the joker. But these, after all, are stylized designs, and it was impossible to tell whether he saw faces or merely patterns. I decided I would show him a volume of cartoons which I had in my briefcase. Here, again, for the most part, he did well. Churchill's cigar, Schnozzle's nose: as soon as he had picked out a key feature, he could identify the face. But cartoons, again, are formal and schematic. It remained to be seen how he would do with real faces, realistically represented.

I turned on the television, keeping the sound off, and found an early Bette Davis film. A love scene was in progress. Dr. P. failed to identify the actress—but this could have been because she had never entered his world. What was more striking was that he failed to identify the expressions on her face or her partner's, though in the course of a single torrid scene these passed from sultry yearning through passion, surprise, disgust, and fury to a melting reconciliation. Dr. P. could make nothing of any of this. He was very unclear as to what was going on, or who was who or even what sex they were. His comments on the scene were positively Martian.

It was just possible that some of his difficulties were associated with the unreality of a celluloid, Hollywood world; and it occurred to me that he might be more successful in identifying faces from his own life. On the walls of the apartment there were photographs of his family, his colleagues, his pupils, himself. I gathered a pile of these together and, with some misgivings, presented them to him. What had been funny, or farcical, in relation to the movie, was tragic in relation to real life. By and large, he recognized nobody: neither his family, nor his colleagues, nor his pupils, nor himself. He recognized a portrait of Einstein because he picked up the characteristic hair and moustache; and the same thing happened with one or two other people. "Ach, Paul!" he said, when shown a portrait of his brother. "That square jaw, those big teeth—I would know Paul anywhere!" But was it Paul he recognized, or one or two of his features, on the basis of which he could make a reasonable guess as to the subject's identity? In the absence of obvious "markers," he was utterly lost. But it was not merely the cognition, the *gnosis*, at fault; there was something radically wrong with the whole way he proceeded. For he approached these faces— even of those near and dear—as if they were abstract puzzles or tests. He did not relate to them, he did not behold. No face was familiar to him, seen as a "thou," being just identified as a set of features, an "it." Thus there was formal, but no trace of personal, gnosis. And with this went his indifference, or blindness, to expression. A face, to us, is a person looking out—we see, as it were, the person through his *persona*, his face. But for Dr. P. there was no *persona* in this sense—no outward *persona*, and no person within.

I had stopped at a florist on my way to his apartment and bought myself an extravagant red rose for my buttonhole. Now I removed this and handed it to him. He took it like a botanist or morphologist given a specimen, not like a person given a flower.

"About six inches in length," he commented. "A convoluted red form with a linear green attachment."

"Yes," I said encouragingly, "and what do you think it is, Dr. P.?"

"Not easy to say." He seemed perplexed. "It lacks the simple symmetry of the Platonic solids, although it may have a higher symmetry of its own. . . . I think this could be an inflorescence or flower."

"Could be?" I queried.

"Could be," he confirmed.

"Smell it," I suggested, and he again looked somewhat puzzled, as if I had asked him to smell a higher symmetry. But he complied courteously and took it to his nose. Now, suddenly, he came to life.

"Beautiful!" he exclaimed. "An early rose. What a heavenly smell!" He started to hum *"Die Rose, die Lillie . . ."* Reality, it seemed, might be conveyed by smell, not by sight.

I tried one final test. It was still a cold day, in early spring, and I had thrown my coat and gloves on the sofa.

"What is this?" I asked, holding up a glove.

"May I examine it?" he asked, and, taking it from me, he proceeded to examine it as he had examined the geometrical shapes.

"A continuous surface," he announced at last, "infolded on itself. It appears to have"—he hesitated—"five outpouchings, if this is the word."

"Yes," I said cautiously. "You have given me a description. Now tell me what it is."

"A container of some sort?"

"Yes," I said, "and what would it contain?"

"It would contain its contents!" said Dr. P., with a laugh. "There are many possibilities. It could be a change purse, for example, for coins of five sizes. It could . . ."

I interrupted the barmy flow. "Does it not look familiar? Do you think it might contain, might fit, a part of your body?"

No light of recognition dawned on his face.*

No child would have the power to see and speak of "a continuous surface . . . infolded on itself," but any child, any infant, would immediately know a glove as a glove, see it as familiar, as going with a hand. Dr. P. didn't. He saw nothing as familiar. Visually, he was lost in a world of lifeless abstractions. Indeed, he did not have a real visual world, as he did not have a real visual self. He could speak about things, but did not see them face-to-face. Hughlings Jackson, discussing patients with aphasia and left-hemisphere lesions, says they have lost "abstract" and "propositional" thought—and compares them with dogs (or, rather, he compares dogs to patients with aphasia). Dr. P., on the other hand, functioned precisely as a machine functions. It wasn't merely that he displayed the same indifference to the visual world as a computer but— even more strikingly—he construed the world as a computer construes it, by means of key features and schematic relationships. The scheme might be identified—in an "Identi-Kit" way—without the reality being grasped at all.

The testing I had done so far told me nothing about Dr. P.'s inner world. Was it possible that his visual memory and imagination were still intact? I asked him to imagine entering one of our local squares from the north side, to walk through it, in imagination or in memory, and tell me the buildings he might pass as he walked. He listed the buildings on his right side, but none of those on his left. I then asked him to imagine entering the square from the south. Again he mentioned only those buildings that were on the right side, although these were the very buildings he had omitted before. Those he had

---

*Later, by accident, he got it on, and exclaimed, "My God, it's a glove!" This was reminiscent of Kurt Goldstein's patient "Lanuti," who could only recognize objects by trying to use them in action.

"seen" internally before were not mentioned now; presumably, they were no longer "seen." It was evident that his difficulties with leftness, his visual field deficits, were as much internal as external, bisecting his visual memory and imagination.

What, at a higher level, of his internal visualization? Thinking of the almost hallucinatory intensity with which Tolstoy visualizes and animates his characters, I questioned Dr. P. about *Anna Karenina*. He could remember incidents without difficulty, had an undiminished grasp of the plot, but completely omitted visual characteristics, visual narrative, and scenes. He remembered the words of the characters but not their faces; and though, when asked, he could quote, with his remarkable and almost verbatim memory, the original visual descriptions, these were, it became apparent, quite empty for him and lacked sensorial, imaginal, or emotional reality. Thus there was an internal agnosia as well.*

But this was only the case, it became clear, with certain sorts of visualization. The visualization of faces and scenes, of visual narrative and drama—this was profoundly impaired, almost absent. But the visualization of *schemata* was preserved, perhaps enhanced. Thus when I engaged him in a game of mental chess, he had no difficulty visualizing the chessboard or the moves—indeed, no difficulty in beating me soundly.

---

*I have often wondered about Helen Keller's visual descriptions, whether these, for all their eloquence, are somehow empty as well. Or whether, by the transference of images from the tactile to the visual, or, yet more extraordinarily, from the verbal and the metaphorical to the sensorial and the visual, she *did* achieve a power of visual imagery, even though her visual cortex had never been stimulated, directly, by the eyes? But in Dr. P.'s case it is precisely the cortex that was damaged, the organic prerequisite of all pictorial imagery. Interestingly and typically he no longer dreamed pictorially—the "message" of the dream being conveyed in nonvisual terms.

Luria said of Zazetsky that he had entirely lost his capacity to play games but that his "vivid imagination" was unimpaired. Zazetsky and Dr. P. lived in worlds which were mirror images of each other. But the saddest difference between them was that Zazetsky, as Luria said, "fought to regain his lost faculties with the indomitable tenacity of the damned," whereas Dr. P. was not fighting, did not know what was lost, did not indeed know that anything was lost. But who was more tragic, or who was more damned—the man who knew it, or the man who did not?

When the examination was over, Mrs. P. called us to the table, where there was coffee and a delicious spread of little cakes. Hungrily, hummingly, Dr. P. started on the cakes. Swiftly, fluently, unthinkingly, melodiously, he pulled the plates towards him and took this and that in a great gurgling stream, an edible song of food, until, suddenly, there came an interruption: a loud, peremptory rat-tat-tat at the door. Startled, taken aback, arrested by the interruption, Dr. P. stopped eating and sat frozen, motionless, at the table, with an indifferent, blind bewilderment on his face. He saw, but no longer saw, the table; no longer perceived it as a table laden with cakes. His wife poured him some coffee: the smell titillated his nose and brought him back to reality. The melody of eating resumed.

How does he do anything? I wondered to myself. What happens when he's dressing, goes to the lavatory, has a bath? I followed his wife into the kitchen and asked her how, for instance, he managed to dress himself. "It's just like the eating," she explained. "I put his usual clothes out, in all the usual places, and he dresses without difficulty, singing to himself. He does everything singing to himself. But if he is interrupted and loses the thread, he comes to a complete stop, doesn't know his clothes—or his own body. He sings all the time—eating songs, dressing songs, bathing songs, everything. He can't do anything unless he makes it a song."

While we were talking my attention was caught by the pictures on the walls.

"Yes," Mrs. P. said, "he was a gifted painter as well as a singer. The School exhibited his pictures every year."

I strolled past them curiously—they were in chronological order. All his earlier work was naturalistic and realistic, with vivid mood and atmosphere, but finely detailed and concrete. Then, years later, they became less vivid, less concrete, less realistic and naturalistic, but far more abstract, even geometrical and cubist. Finally, in the last paintings, the canvasses became nonsense, or nonsense to me—mere chaotic lines and blotches of paint. I commented on this to Mrs. P.

"Ach, you doctors, you're such Philistines!" she exclaimed. "Can you not see *artistic development*—how he renounced the realism of his earlier years, and advanced into abstract, non-representational art?"

"No, that's not it," I said to myself (but forbore to say it to poor Mrs. P.). He had indeed moved from realism to nonrepresentation to the abstract, yet this was not the artist, but the pathology, advancing—advancing towards a profound visual agnosia, in which all powers of representation and imagery, all sense of the concrete, all sense of reality, were being destroyed. This wall of paintings was a tragic pathological exhibit, which belonged to neurology, not art.

And yet, I wondered, was she not partly right? For there is often a struggle, and sometimes, even more interestingly, a collusion between the powers of pathology and creation. Perhaps, in his cubist period, there might have been both artistic and pathological development, colluding to engender an original form; for as he lost the concrete, so he might have gained in the abstract, developing a greater sensitivity to all the structural elements of line, boundary, contour—an almost Picasso-like power to see, and equally depict, those abstract organizations embedded in, and normally lost in, the concrete. . . . Though in the final pictures, I feared, there was only chaos and agnosia.

We returned to the great music room, with the Bösendorfer in the center, and Dr. P. humming the last torte.

"Well, Dr. Sacks," he said to me. "You find me an interesting case, I perceive. Can you tell me what you find wrong, make recommendations?"

"I can't tell you what I find wrong," I replied, "but I'll say what I find right. You are a wonderful musician, and music is your life. What I would prescribe, in a case such as yours, is a life which consists entirely of music. Music has been the center, now make it the whole, of your life."

This was four years ago—I never saw him again, but I often wondered about how he apprehended the world, given his strange loss of image, visuality, and the perfect preservation of a great musicality. I think that music, for him, had taken the place of image. He had no body-image, he had body-music: this is why he could move and act as fluently as he did, but came to a total confused stop if the "inner music" stopped. And equally with the outside, the world . . .*

In *The World as Representation and Will*, Schopenhauer speaks of music as "pure will." How fascinated he would have been by Dr. P., a man who had wholly lost the world as representation, but wholly preserved it as music or will.

And this, mercifully, held to the end—for despite the grad-

---

*Thus, as I learned later from his wife, though he could not recognize his students if they sat still, if they were merely "images," he might suddenly recognize them if they *moved*. "That's Karl," he would cry. "I know his movements, his body-music."

ual advance of his disease (a massive tumor or degenerative process in the visual parts of his brain) Dr. P. lived and taught music to the last days of his life.

POSTSCRIPT

How should one interpret Dr. P.'s peculiar inability to interpret, to judge, a glove as a glove? Manifestly, here, he could not make a cognitive judgment, though he was prolific in the production of cognitive hypotheses. A judgment is intuitive, personal, comprehensive, and concrete—we "see" how things stand, in relation to one another and oneself. It was precisely this setting, this relating, that Dr. P. lacked (though his judging, in all other spheres, was prompt and normal). Was this due to lack of visual information, or faulty processing of visual information? (This would be the explanation given by a classical, schematic neurology.) Or was there something amiss in Dr. P.'s attitude, so that he could not relate what he saw to himself?

These explanations, or modes of explanation, are not mutually exclusive—being in different modes they could coexist, and both be true. And this is acknowledged, implicitly or explicitly, in classical neurology: implicitly, by Donald Macrae, when he finds the explanation of defective schemata, or defective visual processing and integration, inadequate; explicitly, by Kurt Goldstein, when he speaks of "abstract attitude." But abstract attitude, which allows "categorization," also misses the mark with Dr. P.—and, perhaps, with the concept of "judgment" in general. For Dr. P. *had* abstract attitude—indeed, nothing else. And it was precisely this, his absurd abstractness of attitude—absurd because unleavened with anything else—which rendered him incapable of perceiving identity, or particulars, rendered him incapable of judgment.

Neurology and psychology, curiously, though they talk of everything else, almost never talk of "judgment"—and yet it is precisely the downfall of judgment (whether in specific realms, as with Dr. P., or more generally, as in patients with Korsakov's or frontal lobe syndromes) which constitutes the essence of so many neuropsychological disorders. Judgment and identity may be casualties—but neuropsychology never speaks of them.

And yet, whether in a philosophic sense (Kant's sense), or an empirical and evolutionary sense, judgment is the most important faculty we have. An animal, or a man, may get on very well without "abstract attitude" but will speedily perish if deprived of judgment. Judgment must be the *first* faculty of higher life or mind—yet it is ignored, or misinterpreted, by classical (computational) neurology. And if we wonder how such an absurdity can arise, we find it in the assumptions, or the evolution, of neurology itself. For classical neurology (like classical physics) has always been mechanical—from Hughlings Jackson's mechanical analogies to the computer analogies of today.

Of course, the brain *is* a machine and a computer—everything in classical neurology is correct. But our mental processes, which constitute our being and life, are not just abstract and mechanical but personal as well—and, as such, involve not just classifying and categorizing but continual judging and feeling also. If this is missing, we become computer-like, as Dr. P. was. And, by the same token, if we delete feeling and judging, the personal, from the cognitive sciences, we reduce *them* to something as defective as Dr. P.—and we reduce *our* apprehension of the concrete and real.

By a sort of comic and awful analogy, our current cognitive neurology and psychology resemble nothing so much as poor Dr. P.! We need the concrete and real, as he did; and

we fail to see this, as he failed to see it. Our cognitive sciences are themselves suffering from an agnosia essentially
similar to Dr. P.'s. Dr. P. may therefore serve as a warning
and parable—of what happens to a science which eschews the
judgmental, the particular, the personal, and becomes entirely
abstract and computational.

It was always a matter of great regret to me that, owing
to circumstances beyond my control, I was not able to follow his case further, either in the sort of observations and
investigations described, or in ascertaining the actual disease pathology.

One always fears that a case is "unique," especially if it has
such extraordinary features as those of Dr. P. It was, therefore, with a sense of great interest and delight, not unmixed
with relief, that I found, quite by chance—looking through the
periodical *Brain* for 1956—a detailed description of an almost
comically similar case, similar (indeed identical) neuropsychologically and phenomenologically, though the underlying
pathology (an acute head injury) and all personal circumstances were wholly different. The authors speak of their case
as "unique in the documented history of this disorder"—and
evidently experienced, as I did, amazement at their own findings.* The interested reader is referred to the original 1956

---

*Only since the completion of this book have I found that there is, in fact,
a rather extensive literature on visual agnosia in general, and prosopagnosia in particular. In particular I had the great pleasure recently of meeting
Dr. Andrew Kertesz, who has himself published some extremely detailed
studies of patients with such agnosias (see, for example, his 1979 paper
on visual agnosia). Dr. Kertesz mentioned to me a case known to him of
a farmer who had developed prosopagnosia and in consequence could no
longer distinguish the faces of his *cows*, and of another such patient, an
attendant in a Natural History Museum, who mistook his own reflection
for the diorama of an *ape*. As with Dr. P., and as with Macrae and Trolle's
patient, it is especially the animate which is so absurdly misperceived.

paper by Macrae and Trolle, of which I here subjoin a brief paraphrase, with quotations from the original.

Their patient was a young man of 32, who, following a severe automobile accident, was unconscious for three weeks, and "complained, exclusively, of an inability to recognize faces, even those of his wife and children." Not a single face was "familiar" to him, but there were three he could identify; these were workmates—one with an eye-blinking tic, one with a large mole on his cheek, and a third "because he was so tall and thin that no one else was like him." Each of these, Macrae and Trolle bring out, was "recognised solely by the single prominent feature mentioned." In general (like Dr. P.) he recognized familiars only by their voices.

He had difficulty even recognizing himself in a mirror, as Macrae and Trolle describe in detail: "In the early convalescent phase he frequently, especially when shaving, questioned whether the face gazing at him was really his own, and even though he knew it could physically be none other, on several occasions grimaced or stuck out his tongue 'just to make sure.' By carefully studying his face in the mirror he slowly began to recognize it, but 'not in a flash' as in the past—he relied on the hair and facial outline, and on two small moles on his left cheek."

In general he could not recognize objects "at a glance," but would have to seek out, and guess from, one or two features—occasionally his guesses were absurdly wrong. In particular, the authors note, there was difficulty with the *animate*.

On the other hand, simple schematic objects—scissors, watch, key, etc.—presented no difficulties. Macrae and Trolle also note that "His *topographical memory* was strange: the

---

The most important studies of such agnosias, and of visual processing in general, are now being undertaken by A. R. and H. Damasio.

seeming paradox existed that he could find his way from home to hospital and around the hospital, but yet could not name streets *en route* [unlike Dr. P., he also had some aphasia] or appear to visualize the topography."

It was also evident that his visual memories of people, even from long before the accident, were severely impaired—there was memory of conduct, or perhaps a mannerism, but not of visual appearance or face. Similarly, it appeared, when he was questioned closely, that he no longer had visual images in his *dreams*. Thus, as with Dr. P., it was not just visual perception, but visual imagination and memory, the fundamental powers of visual representation, which were essentially damaged in this patient—at least those powers insofar as they pertained to the personal, the familiar, the concrete.

A final, humorous point. Where Dr. P. might mistake his wife for a hat, Macrae's patient, also unable to recognize his wife, needed her to identify herself by a visual *marker*, by "a conspicuous article of clothing, such as a large hat."

2

The Lost Mariner

You have to begin to lose your memory, if only in bits and
pieces, to realize that memory is what makes our lives.
Life without memory is no life at all. . . . Our memory is
our coherence, our reason, our feeling, even our action.
Without it, we are nothing. . . . (I can only wait for the
final amnesia, the one that can erase an entire life, as it
did my mother's.)

—Luis Buñuel

This moving and frightening segment in Buñuel's recently
translated memoirs raises fundamental questions—
clinical, practical, existential, philosophical: what sort of a
life (if any), what sort of a world, what sort of a self, can be
preserved in a man who has lost the greater part of his memory and, with this, his past, and his moorings in time?

It immediately made me think of a patient of mine in
whom these questions are precisely exemplified: charming,
intelligent, memoryless Jimmie G., who was admitted to our
Home for the Aged near New York City early in 1975, with
a cryptic transfer note saying, "Helpless, demented, confused
and disoriented."

Jimmie was a fine-looking man, with a curly bush of grey
hair, a healthy and handsome forty-nine-year-old. He was
cheerful, friendly, and warm.

"Hiya, Doc!" he said. "Nice morning! Do I take this chair
here?" He was a genial soul, very ready to talk and to answer
any questions I asked him. He told me his name and birth
date, and the name of the little town in Connecticut where he
was born. He described it in affectionate detail, even drew me
a map. He spoke of the houses where his family had lived—he
remembered their phone numbers still. He spoke of school
and school days, the friends he'd had, and his special fondness
for mathematics and science. He talked with enthusiasm of
his days in the Navy—he was seventeen, had just graduated
from high school, when he was drafted in 1943. With his good
engineering mind, he was a "natural" for radio and electron-
ics, and after a crash course in Texas found himself assistant
radio operator on a submarine. He remembered the names of
various submarines on which he had served, their missions,
where they were stationed, the names of his shipmates. He
remembered Morse code, and was still fluent in Morse tapping and touch-typing.

A full and interesting early life, remembered vividly, in
detail, with affection. But there, for some reason, his reminiscences stopped. He recalled, and almost relived, his war
days and service, the end of the war, and his thoughts for the
future. He had come to love the Navy, thought he might stay
in it. But with the GI Bill, and support, he felt he might do
best to go to college. His older brother was in accountancy
school and engaged to a girl, a "real beauty" from Oregon.

With recalling, reliving, Jimmie was full of animation; he
did not seem to be speaking of the past but of the present,
and I was very struck by the change of tense in his recollec-
tions as he passed from his school days to his days in the
Navy. He had been using the past tense, but now used the
present—and (it seemed to me) not just the formal or ficti-

tious present tense of recall, but the actual present tense of immediate experience.

A sudden, improbable suspicion seized me.

"What year is this, Mr. G.?" I asked, concealing my perplexity under a casual manner.

"Forty-five, man. What do you mean?" He went on, "We've won the war, FDR's dead, Truman's at the helm. There are great times ahead."

"And you, Jimmie, how old would you be?"

Oddly, uncertainly, he hesitated a moment, as if engaged in calculation.

"Why, I guess I'm nineteen, Doc. I'll be twenty next birthday."

Looking at the grey-haired man before me, I had an impulse for which I have never forgiven myself—it was, or would have been, the height of cruelty had there been any possibility of Jimmie's remembering it.

"Here," I said, and thrust a mirror toward him. "Look in the mirror and tell me what you see. Is that a nineteen-year-old looking out from the mirror?"

He suddenly turned ashen and gripped the sides of the chair. "Jesus Christ," he whispered. "Christ, what's going on? What's happened to me? Is this a nightmare? Am I crazy? Is this a joke?"—and he became frantic, panicked.

"It's okay, Jimmie," I said soothingly. "It's just a mistake. Nothing to worry about. Hey!" I took him to the window. "Isn't this a lovely spring day. See the kids there playing baseball?" He regained his color and started to smile, and I stole away, taking the hateful mirror with me.

Two minutes later I re-entered the room. Jimmie was still standing by the window, gazing with pleasure at the kids playing baseball below. He wheeled around as I opened the door, and his face assumed a cheery expression.

"Hiya, Doc!" he said. "Nice morning! You want to talk to me—do I take this chair here?" There was no sign of recognition on his frank, open face.

"Haven't we met before, Mr. G.?" I asked casually.

"No, I can't say we have. Quite a beard you got there. I wouldn't forget *you*, Doc!"

"Why do you call me 'Doc'?"

"Well, you are a doc, ain't you?"

"Yes, but if you haven't met me, how do you know what I am?"

"You *talk* like a doc. I can *see* you're a doc."

"Well, you're right, I am. I'm the neurologist here."

"Neurologist? Hey, there's something wrong with my nerves? And 'here'—where's 'here'? What is this place anyhow?"

"I was just going to ask you—where do you think you are?"

"I see these beds, and these patients everywhere. Looks like a sort of hospital to me. But hell, what would I be doing in a hospital—and with all these old people, years older than me. I feel good, I'm strong as a bull. Maybe I *work* here. . . . Do I work? What's my job? . . . No, you're shaking your head, I see in your eyes I don't work here. If I don't work here, I've been *put* here. Am I a patient, am I sick and don't know it, Doc? It's crazy, it's scary . . . is it some sort of joke?"

"You don't know what the matter is? You really don't know? You remember telling me about your childhood, growing up in Connecticut, working as a radio operator on submarines? And how your brother is engaged to a girl from Oregon?"

"Hey, you're right. But I didn't tell you that, I never met you before in my life. You must have read all about me in my chart."

"Okay," I said. "I'll tell you a story. A man went to his doctor complaining of memory lapses. The doctor asked him some routine questions, and then said, 'These lapses. What about them?' 'What lapses?' the patient replied."

"So that's my problem," Jimmie laughed. "I kinda thought it was. I do find myself forgetting things, once in a while—things that have just happened. The past is clear, though."

"Will you allow me to examine you, to run over some tests?"

"Sure," he said genially. "Whatever you want."

On intelligence testing he showed excellent ability. He was quick-witted, observant, and logical, and had no difficulty solving complex problems and puzzles—no difficulty, that is, if they could be done quickly. If much time was required, he forgot what he was doing. He was quick and good at tic-tac-toe and checkers, and cunning and aggressive—he easily beat me. But he got lost at chess—the moves were too slow.

Homing in on his memory, I found an extreme and extraordinary loss of recent memory—so that whatever was said or shown to him was apt to be forgotten in a few seconds' time. Thus I laid out my watch, my tie, and my glasses on the desk, covered them, and asked him to remember these. Then, after a minute's chat, I asked him what I had put under the cover. He remembered none of them—or indeed that I had even asked him to remember. I repeated the test, this time getting him to write down the names of the three objects; again he forgot, and when I showed him the paper with his writing on it he was astounded, and said he had no recollection of writing anything down, though he acknowledged that it was his own writing, and then got a faint "echo" of the fact that he had written them down.

He sometimes retained faint memories, some dim echo or sense of familiarity. Thus, five minutes after I had played tic-tac-toe with him, he recollected that "some doctor" had played this with him "a while back"—whether the "while back" was

minutes or months ago he had no idea. He then paused and
said, "It could have been you?" When I said it *was* me, he
seemed amused. This faint amusement and indifference were
very characteristic, as were the involved cogitations to which
he was driven by being so disoriented and lost in time. When I
asked Jimmie the time of the year, he would immediately look
around for some clue—I was careful to remove the calendar
from my desk—and would work out the time of year, roughly,
by looking through the window.

It was not, apparently, that he failed to register in memory,
but that the memory traces were fugitive in the extreme, and
were apt to be effaced within a minute, often less, especially
if there were distracting or competing stimuli, while his intellectual and perceptual powers were preserved, and highly
superior.

Jimmie's scientific knowledge was that of a bright high
school graduate with a penchant for mathematics and science. He was superb at arithmetical (and also algebraic) calculations, but only if they could be done with lightning speed.
If there were many steps, too much time involved, he would
forget where he was, and even the question. He knew the
elements, compared them, and drew the periodic table—but
omitted the transuranic elements.

"Is that complete?" I asked when he'd finished.

"It's complete and up-to-date, sir, as far as I know."

"You wouldn't know any elements beyond uranium?"

"You kidding? There's ninety-two elements, and uranium's
the last."

I paused and flipped through a *National Geographic* on
the table. "Tell me the planets," I said, "and something about
them." Unhesitatingly, confidently, he gave me the planets—
their names, their discovery, their distance from the sun, their
estimated mass, character, and gravity.

"What is this?" I asked, showing him a photo in the magazine I was holding.

"It's the moon," he replied.

"No, it's not," I answered. "It's a picture of the earth taken from the moon."

"Doc, you're kidding! Someone would've had to get a camera up there!"

"Naturally."

"Hell! You're joking—how the hell would you do that?"

Unless he was a consummate actor, a fraud simulating an astonishment he did not feel, this was an utterly convincing demonstration that he was still in the past. His words, his feelings, his innocent wonder, his struggle to make sense of what he saw, were precisely those of an intelligent young man in the 1940s faced with the future, with what had not yet happened, and what was scarcely imaginable. "This more than anything else," I wrote in my notes, "persuades me that his cut-off around 1945 is genuine. . . . What I showed him, and told him, produced the authentic amazement which it would have done in an intelligent young man of the pre-Sputnik era."

I found another photo in the magazine and pushed it over to him.

"That's an aircraft carrier," he said. "Real ultramodern design. I never saw one quite like that."

"What's it called?" I asked.

He glanced down, looked baffled, and said, "The *Nimitz*!"

"Something the matter?"

"The hell there is!" he replied hotly. "I know 'em all by name, and I *don't know* a *Nimitz*. . . . Of course there's an Admiral Nimitz, but I never heard they named a carrier after him."

Angrily he threw the magazine down.

He was becoming fatigued, and somewhat irritable and
anxious, under the continuing pressure of anomaly and contradiction, and their fearful implications, to which he could
not be entirely oblivious. I had already, unthinkingly, pushed
him into panic, and felt it was time to end our session. We
wandered over to the window again and looked down at the
sunlit baseball diamond; as he looked his face relaxed, he
forgot the *Nimitz*, the satellite photo, the other horrors and
hints, and became absorbed in the game below. Then, as a
savory smell drifted up from the dining room, he smacked his
lips, said "Lunch!," smiled, and took his leave.

And I myself was wrung with emotion—it was heartbreak-
ing, it was absurd, it was deeply perplexing, to think of his life
lost in limbo, dissolving.

"He is, as it were," I wrote in my notes, "isolated in a single moment of being, with a moat or lacuna of forgetting all
round him. . . . He is man without a past (or future), stuck
in a constantly changing, meaningless moment." And then,
more prosaically, "The remainder of the neurological examination is entirely normal. Impression: probably Korsakov's
syndrome, due to alcoholic degeneration of the mammillary
bodies." My note was a strange mixture of facts and observations, carefully noted and itemized, with irrepressible medita-
tions on what such problems might "mean," in regard to who
and what and where this poor man was—whether, indeed, one
could speak of an "existence," given so absolute a privation of
memory or continuity.

I kept wondering, in this and later notes—unscientifically—
about "a lost soul," and how one might establish some continuity, some roots, for he was a man without roots, or rooted
only in the remote past.

"Only connect"—but how could he connect, and how could

we help him to connect? What was life without connection? "I may venture to affirm," Hume wrote, "that we are nothing but a bundle or collection of different sensations, which succeed each other with an inconceivable rapidity, and are in a perpetual flux and movement." In some sense, he had been reduced to a "Humean" being—I could not help thinking how fascinated Hume would have been at seeing in Jimmie his own philosophical "chimaera" incarnate, a gruesome reduction of a man to mere disconnected, incoherent flux and change.

Perhaps I could find advice or help in the medical literature—a literature which, for some reason, was largely Russian, from Korsakov's original 1887 thesis about such cases of memory loss, which are still called "Korsakov's syndrome," to A. R. Luria's *Neuropsychology of Memory* (which appeared in translation only a year after I first saw Jimmie). Korsakov wrote in 1887:

Memory of recent events is disturbed almost exclusively; recent impressions apparently disappear soonest, whereas impressions of long ago are recalled properly, so that the patient's ingenuity, his sharpness of wit, and his resourcefulness remain largely unaffected.

To Korsakov's brilliant but spare observations, almost a century of further research has been added—the richest and deepest, by far, being Luria's. And in Luria's account science became poetry, and the pathos of radical lostness was evoked. "Gross disturbances of the organization of impressions of events and their sequence in time can always be observed in such patients," he wrote. "In consequence, they lose their integral experience of time and begin to live in a world of isolated impressions." Further, as Luria noted, the eradication of impressions (and their disorder) might spread backward in time—"in the most serious cases—even to relatively distant events."

Most of Luria's patients, as described in this book, had massive and serious cerebral tumors, which had the same effects as Korsakov's syndrome, but later spread and were often fatal. Luria included no cases of "simple" Korsakov's syndrome, based on the self-limiting destruction that Korsakov described—neuron destruction, produced by alcohol, in the tiny but crucial mammillary bodies, the rest of the brain being perfectly preserved. And so there was no long-term follow-up of Luria's cases.

I had at first been deeply puzzled and dubious, even suspicious, about the apparently sharp cutoff in 1945, a point, a date, which was also symbolically so sharp. I wrote in a subsequent note:

There is a great blank. We do not know what happened then—or subsequently. . . . We must fill in these "missing" years—from his brother, or the Navy, or hospitals he has been to. . . . Could it be that he sustained some massive trauma at this time, some massive cerebral or emotional trauma in combat, in the war, and that *this* may have affected him ever since? . . . Was the war his "high point," the last time he was really alive, and existence since one long anti-climax?*

*In his fascinating oral history *The Good War*, Studs Terkel transcribes countless stories of men and women, especially fighting men, who felt World War II was intensely real, by far the most real and significant time of their lives—and felt everything since as pallid in comparison. Such men tend to dwell on the war and to relive its battles, comradeship, moral certainties and intensity. But this dwelling on the past and relative hebetude towards the present—this emotional dulling of current

We did various tests on him (EEG, brain scans) and found no evidence of massive brain damage, although atrophy of the tiny mammillary bodies would not show up on such tests. We received reports from the Navy indicating that he had remained in the Navy until 1965, and that he was perfectly competent at that time.

Then we turned up a short nasty report from Bellevue Hospital, dated 1971, saying that he was "totally disoriented . . . with an advanced organic brain syndrome, due to alcohol" (cirrhosis had also developed by this time). From Bellevue he was sent to a wretched dump in the Village, a so-called "nursing home" whence he was rescued—lousy, starving—by our Home in 1975.

We located his brother, whom Jimmie always spoke of as being in accountancy school and engaged to a girl from Oregon. In fact, he had married the girl from Oregon, had become a father and grandfather, and been a practicing accountant for thirty years.

Where we had hoped for an abundance of information and feeling from his brother, we received a courteous but somewhat meager letter. It was obvious from reading this— especially reading between the lines—that the brothers had scarcely seen each other since 1943, and gone separate ways, partly through the vicissitudes of location and profession, and partly through deep (though not estranging) differences of temperament. Jimmie, it seemed, had never "settled down," was "happy-go-lucky," and "always a drinker." The Navy, his

---

feeling and memory—is nothing like Jimmie's organic amnesia. I recently had occasion to discuss the question with Terkel. "I've met thousands of men," he told me, "who feel they've just been 'marking time' since '45— but I never met anyone for whom time terminated, like your amnesiac Jimmie."

brother felt, provided a structure, a life, and the real problems started when he left it, in 1965. Without his habitual structure and anchor Jimmie had ceased to work, "gone to pieces," and started to drink heavily. There had been some memory impairment, of the Korsakov type, in the middle and especially the late 1960s, but not so severe that Jimmie couldn't "cope" in his nonchalant fashion. But his drinking grew heavier in 1970.

Around Christmas of that year, his brother understood, he had suddenly "blown his top" and become deliriously excited and confused, and it was at this point he had been taken into Bellevue. During the next month, the excitement and delirium died down, but he was left with deep and bizarre memory lapses, or "deficits," to use the medical jargon. His brother had visited him at this time—they had not met for twenty years—and, to his horror, Jimmie not only failed to recognize him, but said, "Stop joking! You're old enough to be my father. My brother's a young man, just going through accountancy school."

When I received this information, I was more perplexed still: why did Jimmie not remember his later years in the Navy, why did he not recall and organize his memories until 1970? I had not heard then that such patients might have a retrograde amnesia (see Postscript). "I wonder, increasingly," I wrote at this time, "whether there is not an element of hysterical or fugal amnesia—whether he is not in flight from something too awful to recall," and I suggested he be seen by our psychiatrist. Her report was searching and detailed—the examination had included a sodium amytal test, calculated to "release" any memories which might be repressed. She also attempted to hypnotize Jimmie, in the hope of eliciting memories repressed by hysteria—this tends to work well in cases of hysterical amnesia. But it failed because Jimmie could not be hypnotized, not because of any "resistance," but because of his extreme amnesia, which caused him to lose track of what the hypnotist was saying. (Dr. Mark Homonoff, who worked on the amnesia ward at the Boston Veterans Administration hospital, tells me of similar experiences—and of his feeling that this is absolutely characteristic of patients with Korsakov's, as opposed to patients with hysterical amnesia.)

"I have no feeling or evidence," the psychiatrist wrote, "of any hysterical or 'put-on' deficit. He lacks both the means and the motive to make a facade. His memory deficits are organic and permanent and incorrigible, though it is puzzling they should go back so long." Since, she felt, he was "unconcerned . . . manifested no special anxiety . . . constituted no management problem," there was nothing she could offer, or any therapeutic "entrance" or "lever" she could see.

At this point, persuaded that this was, indeed, pure Korsakov's, uncomplicated by other factors, emotional or organic, I wrote to Luria and asked his opinion. He spoke in his reply of his patient Bel, whose amnesia had retroactively eradicated ten years. (He wrote about this patient in *The Neuropsychology of Memory*.) He said he saw no reason why such a retrograde amnesia should not thrust backward decades, or almost a whole lifetime. "I can only wait for the final amnesia," Buñuel writes, "the one that can erase an entire life." But Jimmie's amnesia, for whatever reason, had erased memory and time back to 1945—roughly—and then stopped. Occasionally, he would recall something much later, but the recall was fragmentary and dislocated in time. Once, seeing the word "satellite" in a newspaper headline, he said offhandedly that he'd been involved in a project of satellite tracking while on the ship *Chesapeake Bay*, a memory fragment coming from the early or mid-1960s. But, for all practical purposes, his cutoff point was during the mid- (or late) 1940s, and anything subsequently retrieved was fragmentary, unconnected. This was the case in 1975, and it is still the case now, nine years later.

What could we do? What should we do? "There are no prescriptions," Luria wrote, "in a case like this. Do whatever your ingenuity and your heart suggest. There is little or no hope of any recovery in his memory. But a man does not consist of memory alone. He has feeling, will, sensibilities, moral being—matters of which neuropsychology cannot speak. And it is here, beyond the realm of an impersonal psychology, that you may find ways to touch him, and change him. And the circumstances of your work especially allow this, for you work in a Home, which is like a little world, quite different from the clinics and institutions where I work. Neuropsychologically, there is little or nothing you can do; but in the realm of the Individual, there may be much you can do."

Luria mentioned his patient Kur as manifesting a rare self-awareness, in which hopelessness was mixed with an odd equanimity. "I have no memory of the present," Kur would say. "I do not know what I have just done or from where I have just come. . . . I can recall my past very well, but I have no memory of my present." When asked whether he had ever seen the person testing him, he said, "I cannot say yes or no, I can neither affirm nor deny that I have seen you." This was sometimes the case with Jimmie; and, like Kur, who stayed many months in the same hospital, Jimmie began to form "a sense of familiarity"; he slowly learned his way around the home—the whereabouts of the dining room, his own room, the elevators, the stairs, and in some sense recognized some of the staff, although he confused them, and perhaps had to do so, with people from the past. He soon became fond of the nursing sister in the Home; he recognized her voice, her footfalls, immediately, but would always say that she had been a fellow pupil at his high school, and was greatly surprised when I addressed her as "Sister."

"Gee!" he exclaimed, "the damnedest things happen. I'd never have guessed you'd become a religious, Sister!"

Since he's been at our Home—that is, since early 1975—Jimmie has never been able to identify anyone in it consistently. The only person he truly recognizes is his brother, whenever he visits from Oregon. These meetings are deeply emotional and moving to observe—the only truly emotional meetings Jimmie has. He loves his brother, he recognizes him, but he cannot understand why he looks so old: "Guess some people age fast," he says. Actually his brother looks much younger than his age, and has the sort of face and build that change little with the years. These are true meetings, Jimmie's only connection of past and present, yet they do nothing to provide any sense of history or continuity. If anything they emphasize—at least to his brother, and to others who see them together—that Jimmie still lives, is fossilized, in the past.

All of us, at first, had high hopes of helping Jimmie—he was so personable, so likable, so quick and intelligent, it was difficult to believe that he might be beyond help. But none of us had ever encountered, even imagined, such a power of amnesia, the possibility of a pit into which everything, every experience, every event, would fathomlessly drop, a bottomless memory-hole that would engulf the whole world.

I suggested, when I first saw him, that he should keep a diary, and be encouraged to keep notes every day of his experiences, his feelings, thoughts, memories, reflections. These attempts were foiled, at first, by his continually losing the diary: it had to be attached to him—somehow. But this too failed to work; he dutifully kept a brief daily notebook but could not recognize his earlier entries in it. He does recognize his own writing, and style, and is always astounded to find that he wrote something the day before.

Astounded—and indifferent—for he was a man who, in effect, had no "day before." His entries remained unconnected and unconnecting and had no power to provide any sense of time or continuity. Moreover, they were trivial—"Eggs for breakfast," "Watched ballgame on TV"—and never touched the depths. But were there depths in this unmemoried man, depths of an abiding feeling and thinking, or had he been reduced to a sort of Humean drivel, a mere succession of unrelated impressions and events?

Jimmie both was and wasn't aware of this deep, tragic loss in himself, loss *of* himself. (If a man has lost a leg or an eye, he knows he has lost a leg or an eye; but if he has lost a self—himself—he cannot know it, because he is no longer there to know it.) Therefore I could not question him intellectually about such matters.

He had originally professed bewilderment at finding himself amid patients, when, as he said, he himself didn't feel ill. But what, we wondered, did he feel? He was strongly built and fit, he had a sort of animal strength and energy, but also a strange inertia, passivity, and (as everyone remarked) "unconcern"; he gave all of us an overwhelming sense of "something missing," although this, if he realized it, was itself accepted with an odd "unconcern." One day I asked him not about his memory, or past, but about the simplest and most elemental feelings of all:

"How do you feel?"

"How do I feel," he repeated, and scratched his head. "I cannot say I feel ill. But I cannot say I feel well. I cannot say I feel anything at all."

"Are you miserable?" I continued.

"Can't say I am."

"Do you enjoy life?"

"I can't say I do. . . ."

I hesitated, fearing that I was going too far, that I might be stripping a man down to some hidden, unacknowledgeable, unbearable despair.

"You don't enjoy life," I repeated, hesitating somewhat. "How then *do* you feel about life?"

"I can't say that I feel anything at all."

"You feel alive though?"

"Feel alive? Not really. I haven't felt alive for a very long time."

His face wore a look of infinite sadness and resignation.

Later, having noted his aptitude for, and pleasure in, quick games and puzzles, and their power to "hold" him, at least while they lasted, and to allow, for a while, a sense of companionship and competition—he had not complained of loneliness, but he looked so alone; he never expressed sadness, but he looked so sad—I suggested he be brought into our recreation programs at the Home. This worked better, better than the diary. He would become keenly and briefly involved in games, but soon they ceased to offer any challenge: he solved all the puzzles, and could solve them easily; and he was far better and sharper than anyone else at games. And as he found this out, he grew fretful and restless again, and wandered the corridors, uneasy and bored and with a sense of indignity—games and puzzles were for children, a diversion. Clearly, passionately, he wanted something to do: he wanted to do, to be, to feel—and could not; he wanted sense, he wanted purpose—in Freud's words, "Work and Love."

Could he do "ordinary" work? He had "gone to pieces," his brother said, when he ceased to work in 1965. He had two striking skills—Morse code and touch-typing. We could not use Morse, unless we invented a use; but good typing we could use, if he could recover his old skills—and this would be real work, not just a game. Jimmie soon did recover his old skill and came to type very quickly—he could not do it slowly—and found in this some of the challenge and satisfaction of a job. But still this was superficial tapping and typing; it was trivial, it did not reach to the depths. And what he typed, he typed mechanically—he could not hold the thought—the short sentences following one another in a meaningless order.

One tended to speak of him, instinctively, as a spiritual casualty—a "lost soul": was it possible that he had really been "desouled" by a disease? "Do you think he *has* a soul?" I once asked the Sisters. They were outraged by my question, but could see why I asked it. "Watch Jimmie in chapel," they said, "and judge for yourself."

I did, and I was moved, profoundly moved and impressed, because I saw here an intensity and steadiness of attention and concentration that I had never seen before in him or conceived him capable of. I watched him kneel and take the Sacrament on his tongue, and could not doubt the fullness and totality of Communion, the perfect alignment of his spirit with the spirit of the Mass. Fully, intensely, quietly, in the quietude of absolute concentration and attention, he entered and partook of the Holy Communion. He was wholly held, absorbed, by a feeling. There was no forgetting, no Korsakov's then, nor did it seem possible or imaginable that there should be; for he was no longer at the mercy of a faulty and fallible mechanism—that of meaningless sequences and memory traces—but was absorbed in an act, an act of his whole being, which carried feeling and meaning in an organic continuity and unity, a continuity and unity so seamless it could not permit any break.

Clearly Jimmie found himself, found continuity and reality, in the absoluteness of spiritual attention and act. The Sisters were right—he did find his soul here. And so was Luria, whose words now came back to me: "A man does not consist of memory alone. He has feeling, will, sensibility, moral being. . . . It is here . . . you may touch him, and see a profound change." Memory, mental activity, mind alone, could not hold him; but moral attention and action could hold him completely.

But perhaps "moral" was too narrow a word—for the aesthetic and dramatic were equally involved. Seeing Jimmie in the chapel opened my eyes to other realms where the soul is called on, and held, and stilled, in attention and communion. The same depth of absorption and attention was to be seen in relation to music and art: he had no difficulty, I noticed, "following" music or simple dramas, for every moment in music and art refers to, contains, other moments. He liked gardening, and had taken over some of the work in our garden. At first he greeted the garden each day as new, but for some reason this had become more familiar to him than the inside of the Home. He almost never got lost or disoriented in the garden now; he patterned it, I think, on loved and remembered gardens from his youth in Connecticut.

Jimmie, who was so lost in extensional "spatial" time, was perfectly organized in Bergsonian "intentional" time; what was fugitive, unsustainable, as formal structure, was perfectly stable, perfectly held, as art or will. Moreover, there was something that endured and survived. If Jimmie was briefly "held" by a task or puzzle or game or calculation, held in the purely mental challenge of these, he would fall apart as soon as they were done, into the abyss of his nothingness, his amnesia. But if he was held in emotional and spiritual attention—in the contemplation of nature or art, in listening to music, in taking part in the Mass in chapel—the attention, its "mood," its quietude, would persist for a while, and there would be in him a pensiveness and peace we rarely, if ever, saw during the rest of his life at the Home.

I have known Jimmie now for nine years—and neuropsychologically, he has not changed in the least. He still has the severest, most devastating Korsakov's, cannot remember isolated items for more than a few seconds, and has a dense amnesia going back to 1945. But humanly, spiritually, he is at times a different man altogether—no longer fluttering, restless, bored, and lost, but deeply attentive to the beauty and soul of the world, rich in all the Kierkegaardian categories— and aesthetic, the moral, the religious, the dramatic. I had wondered, when I first met him, if he was not condemned to a sort of "Humean" froth, a meaningless fluttering on the surface of life, and whether there was any way of transcending the incoherence of his Humean disease. Empirical science told me there was not—but empirical science, empiricism, takes no account of the soul, no account of what constitutes and determines personal being. Perhaps there is a philosophical as well as a clinical lesson here: that in Korsakov's, or dementia, or other such catastrophes, however great the organic damage and Humean dissolution, there remains the undiminished possibility of reintegration by art, by communion, by touching the human spirit: and this can be preserved in what seems at first a hopeless state of neurological devastation.

POSTSCRIPT

I know now that retrograde amnesia, to some degree, is very common, if not universal, in cases of Korsakov's. The classical Korsakov's syndrome—a profound and permanent, but "pure," devastation of memory caused by alcoholic destruction of the mammillary bodies—is rare, even among very heavy drinkers. One may, of course, see Korsakov's syndrome with other pathologies, as in Luria's patients with tumors. A particularly fascinating case of an acute (and mercifully transient) Korsakov's syndrome has been well described only very recently in the so-called transient global amnesia (TGA) which may occur with migraines, head injuries or impaired blood supply to the brain. Here, for a few minutes or hours, a severe and singular amnesia may occur, even though the patient may continue to drive a car, or, perhaps, to carry on medical or editorial duties, in a mechanical way. But under this fluency lies a profound amnesia—every sentence uttered being forgotten as soon as it is said, everything forgotten within a few minutes of being seen, though long-established memories and routines may be perfectly preserved. (Some remarkable videotapes of patients *during* TGAs have recently [1986] been made by Dr. John Hodges, of Oxford.)

Further, there may be a profound retrograde amnesia in such cases. My colleague Dr. Leon Protass tells me of a case he saw recently, in which a highly intelligent man was unable for some hours to remember his wife or children, to remember that he had a wife or children. In effect, he lost thirty years of his life—though, fortunately, for only a few hours. Recovery from such attacks is prompt and complete—yet they are, in a sense, the most horrifying of "little strokes" in their power absolutely to annul or obliterate decades of richly lived, richly achieving, richly memoried life. The horror, typically, is only felt by others—the patient, unaware, amnesiac for his amnesia, may continue what he is doing, quite unconcerned, and only discover later that he lost not only a day (as is common with ordinary alcoholic "blackouts"), but half a lifetime, and never knew it. The fact that one can lose the greater part of a lifetime has peculiar, uncanny horror.

In adulthood, life, higher life, may be brought to a premature end by strokes, senility, brain injuries, etc., but there usually remains the consciousness of life lived, of one's past. This is usually felt as a sort of compensation: "At least I lived fully, tasting life to the full, before I was brain injured, stricken, etc." This sense of "the life lived before," which may be either a consolation or a torment, is precisely what is taken away in retrograde amnesia. The "final amnesia, the one that can erase an entire life" that Buñuel speaks of may occur, perhaps, in a terminal dementia, but not, in my experience, suddenly, in consequence of a stroke. But there is a different yet comparable sort of amnesia, which can occur suddenly—different in that it is not "global" but "modality-specific."

Thus, in one patient under my care, a sudden thrombosis in the posterior circulation of the brain caused the immediate death of the visual parts of his brain. Forthwith this patient became completely blind—but did not know it. He looked blind—but he made no complaints. Questioning and testing showed, beyond doubt, that not only was he centrally or "cortically" blind, but he had lost all visual images and memories, lost them totally—yet had no sense of any loss. Indeed, he had lost the very idea of seeing—and was not only unable to describe anything visually, but bewildered when I used words such as "seeing" and "light." He had become, in essence, a non-visual being. His entire lifetime of seeing, of visuality, had, in effect, been stolen. His whole visual life had, indeed, been erased—and erased permanently in the instant of his stroke. Such a visual amnesia, and (so to speak) blindness to the blindness, amnesia for the amnesia, is in effect a "total" Korsakov's, confined to visuality.

A still more limited, but nonetheless total, amnesia may be displayed with regard to particular forms of perception, as in the previous chapter, "The Man Who Mistook His Wife for a Hat." There, there was an absolute "prosopagnosia," or agnosia for faces. Dr. P. was not only unable to recognize faces but unable to imagine or remember any faces—he had indeed lost the very idea of a "face," as my more afflicted patient had lost the very ideas of "seeing" or "light." Such syndromes were described by Gabriel Anton in the 1890s. But the implication of these syndromes—Korsakov's and Anton's—what they entail and must entail for the world, the lives, the identities of affected patients, has been scarcely touched on even to this day.

In Jimmie's case, we had sometimes wondered how he might respond if taken back to his hometown—in effect, to his pre-amnesia days—but the little town in Connecticut had become a booming city with the years. Later I did have occasion to find out what might happen in such circumstances, though this was with another patient with Korsakov's, Stephen R., who had become acutely ill in 1980 and whose retrograde amnesia went back only two years or so. With this patient, who also had severe seizures, spasticity, and other problems necessitating inpatient care, rare weekend visits to his home revealed a poignant situation. In hospital he could recognize nobody and nothing, and was in an almost ceaseless frenzy of disorientation. But when his wife took him home, to his house which was in effect a "time capsule" of his pre-amnesia days, he felt instantly at home. He recognized everything, tapped the barometer, checked the thermostat, took his favorite armchair, as he used to do. He spoke of neighbors, shops, the local pub, a nearby cinema, as they had been in the midseventies. He was distressed and puzzled if the smallest changes were made in the house. ("You changed the curtains today!" he once expostulated to his wife. "How come? So suddenly? They were green this morning." But they had not been green since 1978.) He recognized most of the neighboring houses and shops—they had changed little between 1978 and 1983—but was bewildered by the "replacement" of the cinema ("How could they tear it down and put up a supermarket *overnight*?"). He recognized friends and neighbors—but found them oddly older than he expected ("Old so-and-so! He's really showing his age. Never noticed it before. How come everyone's showing their age today?"). But the real poignancy, the horror, would occur when his wife brought him back— brought him, in a fantastic and unaccountable manner (so he felt), to a strange home he had never seen, full of strangers, and then left him. "What are you doing?" he would scream, terrified and confused. "What in the hell *is* this place? What the hell's going on?" These scenes were almost unbearable to watch, and must have seemed like madness, or nightmare, to the patient. Mercifully perhaps he would forget them within a couple of minutes.

Such patients, fossilized in the past, can only be at home, oriented, in the past. Time, for them, has come to a stop. I hear Stephen R. screaming with terror and confusion when he returns—screaming for a past which no longer exists. But what can we do? Can we create a time capsule, a fiction? Never have I known a patient so confronted, so tormented, by anachronism, unless it was "Rose R." of *Awakenings* (see "Incontinent Nostalgia").

Jimmie has reached a sort of calm; William (Chapter 12) continually confabulates; but Stephen R. has a gaping time-wound, an agony that will never heal.*

*After writing and publishing this history I embarked with Dr. Elkhonon Goldberg—a pupil of Luria and editor of the original Russian edition of *The Neuropsychology of Memory*—on a close and systematic neuropsychological study of this patient. Dr. Goldberg has presented some of the preliminary findings at conferences, and we hope in due course to publish a full account.

A deeply moving and extraordinary film about a patient with a profound amnesia (*Prisoner of Consciousness*), made by Dr. Jonathan Miller, has just been shown in England (September 1986). A film has also been made (by Hilary Lawson) with a prosopagnosic patient with many similarities to Dr. P. Such films are crucial to assist the imagination: "What *can* be shown *cannot* be said."

3

The Disembodied Lady

The aspects of things that are most important for us are hidden because of their simplicity and familiarity. (One is unable to notice something because it is always before one's eyes.) The real foundations of his enquiry do not strike a man at all.

—Ludwig Wittgenstein

W hat Wittgenstein writes here, of epistemology, might apply to aspects of one's physiology and psychology— especially in regard to what Charles Sherrington once called "our secret sense, our sixth sense"—that continuous but unconscious sensory flow from the movable parts of our body (muscles, tendons, joints), by which their position and tone and motion are continually monitored and adjusted, but in a way which is hidden from us because it is automatic and unconscious.

Our other senses—the five senses—are open and obvious; but this—our hidden sense—had to be discovered, as it was, by Sherrington, in the 1890s. He named it "proprioception," to distinguish it from "exteroception" and "interoception," and, additionally, because of its indispensability for our sense of *ourselves*; for it is only by courtesy of proprioception, so to speak, that we feel our bodies as proper to us, as our "property," as our own.

What is more important for us, at an elemental level, than the control, the owning and operation, of our own physical selves? And yet it is so automatic, so familiar, we never give it a thought.

Jonathan Miller produced a beautiful television series, *The Body in Question*, but the body, normally, is never in question: our bodies are beyond question, or perhaps beneath question—they are simply, unquestionably, there. This unquestionability of the body, its certainty, is, for Wittgenstein, the start and basis of all knowledge and certainty. Thus in his last book, *On Certainty*, he opens by saying: "If you do know that *here is one hand*, we'll grant you all the rest." But then, in the same breath, on the same opening page: "What we can ask is whether it can make sense to doubt it . . ." and, a little later, "Can I doubt it? Grounds for *doubt* are lacking!"

Indeed, his book might be titled *On Doubt*, for it is marked by doubting, no less than affirming. Specifically, he wonders—and one in turn may wonder whether these thoughts were perhaps incited by his working with patients in a hospital, in the war—whether there might be situations or conditions which take away the certainty of the body, which do give one grounds to doubt one's body, perhaps indeed to lose one's entire body in total doubt. This thought seems to haunt his last book like a nightmare.

Christina was a strapping young woman of 27, given to hockey and riding, self-assured, robust, in body and mind. She had two young children and worked as a computer programmer at home. She was intelligent and cultivated, fond of the ballet, and of the Lakeland poets (but not, I would think, of Wittgenstein). She had an active, full life—had scarcely known a day's illness. Somewhat to her surprise, after an attack of abdominal pain, she was found to have gallstones, and removal of the gallbladder was advised.

She was admitted to hospital three days before the operation date and placed on an antibiotic for microbial prophylaxis. This was purely routine, a precaution, no complications of any sort being expected at all. Christina understood this, and being a sensible soul had no great anxieties.

The day before surgery Christina, not usually given to fancies or dreams, had a disturbing dream of peculiar intensity. She was swaying wildly, in her dream, very unsteady on her feet, could hardly feel the ground beneath her, could hardly feel anything in her hands, found them flailing to and fro, kept dropping whatever she picked up.

She was distressed by this dream. ("I never had one like it," she said. "I can't get it out of my mind.")—so distressed that we requested an opinion from the psychiatrist. "Pre-operative anxiety," he said. "Quite natural, we see it all the time."

But later that day *the dream came true*. Christina did find herself very unsteady on her feet, with awkward flailing movements, and dropping things from her hands.

The psychiatrist was again called—he seemed vexed at the call, but also, momentarily, uncertain and bewildered. "Anxiety hysteria," he now snapped, in a dismissive tone. "Typical conversion symptoms—you see them all the while."

But the day of surgery Christina was still worse. Standing was impossible—unless she looked down at her feet. She could hold nothing in her hands, and they "wandered"—unless she kept an eye on them. When she reached out for something or tried to feed herself, her hands would miss or overshoot wildly, as if some essential control or coordination was gone.

She could scarcely even sit up—her body "gave way." Her face was oddly expressionless and slack, her jaw fell open, even her vocal posture was gone.

"Something awful's happened," she mouthed, in a ghostly flat voice. "I can't feel my body. I feel weird—disembodied."

This was an amazing thing to hear, confounded, confounding. "Disembodied"—was she crazy? But what of her physical state then? The collapse of tone and muscle posture, from top to toe; the wandering of her hands, which she seemed unaware of; the flailing and overshooting, as if she were receiving no information from the periphery, as if the control loops for tone and movement had catastrophically broken down.

"It's a strange statement," I said to the residents. "It's almost impossible to imagine what might provoke such a statement."

"But it's hysteria, Dr. Sacks—didn't the psychiatrist say so?"

"Yes, he did. But have you ever seen a hysteria like this? Think phenomenologically—take what you see as genuine phenomenon, in which her state-of-body and state-of-mind are not fictions, but a psychophysical whole. Could anything give such a picture of undermined body and mind?

"I'm not testing you," I added. "I'm as bewildered as you are. I've never seen or imagined anything quite like this before."

I thought, and they thought, we thought together.

"Could it be a biparietal syndrome?" one of them asked.

"It's an 'as if,'" I answered: "*as if* the parietal lobes were not getting their usual sensory information. Let's *do* some sensory testing—and test parietal lobe function, too."

We did so, and a picture began to emerge. There seemed to be a very profound, almost total, proprioceptive deficit, going from the tips of her toes to her head—the parietal lobes were working, *but had nothing to work with*. Christina might have hysteria, but she had a great deal more, of a sort which none of us had ever seen or conceived before. We put in an emergency call now, not to the psychiatrist, but to the physical medicine specialist, the physiatrist.

He arrived promptly, responding to the urgency of the

call. He opened his eyes very wide when he saw Christina; he examined her swiftly and comprehensively, and then proceeded to electrical tests of nerve and muscle function. "This is quite extraordinary," he said. "I have never seen or read about anything like this before. She has lost all proprioception—you're right—from top to toe. She has no muscle or tendon or joint sense whatever. There is slight loss of other sensory modalities—to light touch, temperature, and pain, and slight involvement of the motor fibers, too. But it is predominantly position-sense—proprioception—which has sustained such damage."

"What's the cause?" we asked.

"You're the neurologists. You find out."

By afternoon, Christina was still worse. She lay motionless and toneless; even her breathing was shallow. Her situation was grave—we thought of a respirator—as well as strange.

The picture revealed by spinal tap was one of an acute polyneuritis, but a polyneuritis of a most exceptional type: not like Guillain-Barré syndrome, with its overwhelming motor involvement, but a purely (or almost purely) sensory neuritis, affecting the sensory roots of spinal and cranial nerves throughout the neuraxis.*

Operation was deferred; it would have been madness at this time. Much more pressing were the questions: Will she survive? What can we do?

"What's the verdict?" Christina asked, with a faint voice and fainter smile, after we had checked her spinal fluid.

---

*Such sensory polyneuropathies occur but are rare. What was unique in Christina's case, to the best of our knowledge at the time (this was in 1977), was the extraordinary selectivity displayed, so that proprioceptive fibers, and these only, bore the brunt of the damage. But see Sterman (1979).

"You've got this inflammation, this neuritis, . . ." we began, and told her all we knew. When we forgot something, or hedged, her clear questions brought us back.

"Will it get better?" she demanded. We looked at each other, and at her: "We have no idea."

The sense of the body, I told her, is given by three things: vision, balance organs (the vestibular system), and proprioception—which she'd lost. Normally all of these worked together. If one failed, the others could compensate, or substitute—to a degree. In particular, I told her of my patient Mr. MacGregor, who, unable to employ his balance organs, used his eyes instead (see "On the Level"). And of patients with neurosyphilis, *tabes dorsalis*, who had similar symptoms, but confined to the legs—and how they too had to compensate by use of their eyes (see "Positional Phantoms" in Chapter 6). And how, if one asked such a patient to move his legs, he was apt to say, "Sure, Doc, as soon as I find them."

Christina listened closely, with a sort of desperate attention.

"What I must do then," she said slowly, "is use vision, use my eyes, in every situation where I used—what do you call it?—proprioception before. I've already noticed," she added, musingly, "that I may 'lose' my arms. I think they're one place, and I find they're another. This 'proprioception' is like the eyes of the body, the way the body sees itself. And if it goes, as it's gone with me, *it's like the body's blind*. My body can't 'see' itself if it's lost its eyes, right? So *I* have to watch it—be its eyes. Right?"

"Right," I said, "right. You could be a physiologist."

"I'll *have* to be a sort of physiologist," she rejoined, "because my physiology has gone wrong, and may never *naturally* go right."

It was as well that Christina showed such strength of mind from the start, for though the acute inflammation subsided, and her spinal fluid returned to normal, the damage it did to her proprioceptive fibers persisted—so that there was no neurological recovery a week, or a year, later. Indeed, there has been none in the eight years that have now passed—though she has been able to lead a life, a sort of life, through accommodations and adjustments of every sort, emotional and moral no less than neurological.

That first week Christina did nothing, lay passively, scarcely ate. She was in a state of utter shock, horror and despair. What sort of a life would it be, if there was not natural recovery? What sort of a life, every move made by artifice? What sort of a life, above all, if she felt disembodied?

Then life reasserted itself, as it will, and Christina started to move. She could at first do nothing without using her eyes, and collapsed in a helpless heap the moment she closed them. She had, at first, to monitor herself by vision, looking carefully at each part of her body as it moved, using an almost painful conscientiousness and care. Her movements, consciously monitored and regulated, were at first clumsy, artificial, in the highest degree. But then—and here both of us found ourselves most happily surprised, by the power of an ever-increasing, daily increasing, automatism—then her movements started to appear more delicately modulated, more graceful, more natural (though still wholly dependent on use of the eyes).

Increasingly now, week by week, the normal, unconscious feedback of proprioception was being replaced by an equally unconscious feedback by vision, by visual automatism and reflexes increasingly integrated and fluent. Was it possible, too, that something more fundamental was happening? That the brain's visual model of the body, or body image—normally rather feeble (it is, of course, absent in the blind), and normally subsidiary to the proprioceptive body model—was it possible that *this*, now the proprioceptive body model was lost, was gaining, by way of compensation or substitution, an enhanced, exceptional, extraordinary force? And to this might be added a compensatory enhancement of the vestibular body model or body image, too . . . both to an extent which was more than we had expected or hoped for.*

Whether or not there was increased use of vestibular feedback, there was certainly increased use of her ears—auditory feedback. Normally this is subsidiary, and rather unimportant in speaking—our speech remains normal if we are deaf from a head cold, and some of the congenitally deaf may be able to acquire virtually perfect speech. For the modulation of speech is normally proprioceptive, governed by inflowing impulses from all our vocal organs. Christina had lost this normal inflow, this afference, and lost her normal proprioceptive vocal tone and posture, and therefore had to use her ears, auditory feedback, instead.

Besides these new, compensatory forms of feedback, Christina also started to develop—it was deliberate and conscious in the first place, but gradually became unconscious and automatic—various forms of new and compensatory "feed-

---

*Contrast the fascinating case described by James Purdon Martin in *The Basal Ganglia and Posture*: "This patient, in spite of years of physiotherapy and training, has never regained the ability to walk in any normal manner. His greatest difficulty is in starting to walk and in propelling himself forward. . . . He is also unable to rise from a chair. He cannot crawl or place himself in the all-fours posture. When standing or walking he is entirely dependent on vision and falls down if he closes his eyes. At first he was unable to maintain his position on an ordinary chair when he closed his eyes, but he has gradually acquired the ability to do this."

forward" (in all this she was assisted by an immensely understanding and resourceful rehabilitative staff).

Thus at the time of her catastrophe, and for about a month
afterwards, Christina remained as floppy as a ragdoll, unable
even to sit up. But three months later, I was startled to see her
sitting very finely—too finely, statuesquely, like a dancer in
mid-pose. And soon I saw that her sitting was, indeed, a pose,
consciously or automatically adopted and sustained, a sort of
forced or willful or histrionic posture, to make up for the continuing lack of any genuine, natural posture. Nature having
failed, she took to "artifice," but the artifice was suggested by
nature, and soon became "second nature."

Similarly with her voice—she had at first been almost mute.
This too was projected, as to an audience from a stage. It *was*
a stagey, theatrical voice—not because of any histrionism,
or perversion of motive, but because there was still no natural vocal posture. And with her face, too—this still tended to
remain somewhat flat and expressionless (though her inner
emotions were of full and normal intensity), due to lack of
proprioceptive facial tone and posture, unless she used an
artificial enhancement of expression (as patients with aphasia
may adopt exaggerated emphases and inflections).*

But all these measures were, at best, partial. They made
life possible—they did not make it normal. Christina learned
to walk, to take public transport, to conduct the usual business of life—but only with the exercise of great vigilance, and
strange ways of doing things—ways which might break down

---

*Purdon Martin, almost alone of contemporary neurologists, would often
speak of facial and vocal "posture" and their basis, finally, in proprioceptive integrity. He was greatly intrigued when I told him about Christina
and showed him some films and tapes of her—many of the suggestions
and formulations here are, in fact, his.

if her attention was diverted. Thus if she was eating while she was talking, or if her attention was elsewhere, she would grip the knife and fork with painful force—her nails and fingertips would go bloodless with pressure; but if there was any lessening of the painful pressure, she might nervelessly drop them straightaway—there was no in-between, no modulation, whatever.

Thus, although there was not a trace of neurological recovery (recovery from the anatomical damage to nerve fibers), there was, with the help of intensive and varied therapy—she remained in hospital, on the rehabilitation ward, for almost a year—a very considerable functional recovery, i.e., the ability to function using various substitutions and other such tricks. It became possible, finally, for Christina to leave hospital, go home, rejoin her children. She was able to return to her home computer terminal, which she now learned to operate with extraordinary skill and efficiency, considering that everything had to be done by vision, not feel. She had learned to operate— but how did she feel? Had the substitutions dispersed the disembodied sense she first spoke of?

The answer is—not in the least. She continues to feel, with the continuing loss of proprioception, that her body is dead, not-real, not-hers—she cannot appropriate it to herself. She can find no words for this state and can only use analogies derived from other senses: "I feel my body is blind and deaf to itself. . . . It has no sense of itself"—these are her own words. She has no words, no direct words, to describe this bereftness, this sensory darkness (or silence) akin to blindness or deafness. She has no words, and we lack words too. And society lacks words, and sympathy, for such states. The blind, at least, are treated with solicitude—we can imagine their state, and we treat them accordingly. But when Christina, painfully,

clumsily, mounts a bus, she receives nothing but uncomprehending and angry snarls: "What's wrong with you, lady? Are
you blind—or blind-drunk?" What can she answer—"I have no
proprioception"? The lack of social support and sympathy is
an additional trial: disabled, but with the nature of her dis-
ability not clear—she is not, after all, manifestly blind or paralyzed, manifestly anything—she tends to be treated as a phony
or a fool. This is what happens to those with disorders of the
hidden senses (it happens also to patients who have vestibular
impairment, or who have been labyrinthectomized).

Christina is condemned to live in an indescribable, unimaginable realm—though "non-realm," "nothingness," might be
better words for it. At times she breaks down—not in public,
but with me: "If only I could *feel*!" she cries. "But I've forgotten what it's like. . . . I *was* normal, wasn't I? I *did* move like
everyone else?"

"Yes, of course."

"There's no 'of course.' I can't believe it. I want proof."

I show her a home movie of herself with her children, taken
just a few weeks before her polyneuritis.

"Yes, of course, that's me!" Christina smiles, and then
cries: "But I can't identify with that graceful girl anymore!
She's gone, I can't remember her, *I can't even imagine her*.
It's like something's been scooped right out of me, right at the
center . . . that's what they do with frogs, isn't it? They scoop
out the center, the spinal cord, they *pith* them. . . . That's
what I am, *pithed*, like a frog. . . . Step right up, come and
see Chris, the first pithed human being. She has no proprioception, no sense of herself—disembodied Chris, the pithed
girl!" She laughs wildly, with an edge of hysteria. I calm her—
"Come now!"—while thinking, "Is she right?"

For, in some sense, she *is* "pithed," disembodied, a sort of wraith. She has lost, with her sense of proprioception, the fundamental, organic mooring of identity—at least of that corporeal identity, or "body-ego," which Freud sees as the basis of self: "The ego is first and foremost a body-ego." Some such depersonalization or derealization must always occur, when there are deep disturbances of body perception or body image. Weir Mitchell saw this, and incomparably described it, when he was working with amputees and nerve-damaged patients in the American Civil War—and in a famous, quasi-fictionalized account, but still the best, phenomenologically most accurate, account we have, said (through the mouth of his physician-patient, George Dedlow):

> I found to my horror that at times I was less conscious of myself, of my own existence, than used to be the case. This sensation was so novel that at first it quite bewildered me. I felt like asking someone constantly if I were really George Dedlow or not; but, well aware of how absurd I should seem after such a question, I refrained from speaking of my case, and strove more keenly to analyze my feelings. At times the conviction of my want of being myself was overwhelming and most painful. It was, as well as I can describe it, a deficiency in the egoistic sentiment of individuality.

For Christina there is this general feeling—this "deficiency in the egoistic sentiment of individuality"—which has become less with accommodation, with the passage of time. And there is this specific, organically based, feeling of disembodiedness, which remains as severe and uncanny as the day she first felt it. This is also felt, for example, by those who have high transections of the spinal cord—but they of course, are paralyzed; whereas Christina, though "bodiless," is up and about.

There are brief, partial reprieves, when her skin is stimulated. She goes out when she can, she loves open cars, where she can feel the wind on her body and face (superficial sensation, light touch, is only slightly impaired). "It's wonderful," she says. "I feel the wind on my arms and face, and then I know, faintly, I *have* arms and a face. It's not the real thing, but it's something—it lifts this horrible, dead veil for a while."

But her situation is, and remains, a Wittgensteinian one. She does not know "Here is one hand"—her loss of proprioception, her deafferentation, has deprived her of her existential, her epistemic, basis—and nothing she can do or think will alter this fact. She cannot be certain of her body—what would Wittgenstein have said, in her position?

In an extraordinary way, she has both succeeded and failed. She has succeeded in operating, but not in being. She has succeeded to an almost incredible extent in all the accommodations that will, courage, tenacity, independence and the plasticity of the senses and the nervous system will permit. She has faced, she faces, an unprecedented situation, has battled against unimaginable difficulties and odds, and has survived as an indomitable, impressive human being. She is one of those unsung heroes, or heroines, of neurological affliction.

But still and forever she remains defective and defeated. Not all the spirit and ingenuity in the world, not all the substitutions or compensations the nervous system allows, can alter in the least her continuing and absolute loss of proprioception—that vital sixth sense without which a body must remain unreal, unpossessed.

Poor Christina is "pithed" in 1985 as she was eight years ago and will remain so for the rest of her life. Her life is unprecedented. She is, so far as I know, the first of her kind, the first "disembodied" human being.

POSTSCRIPT

Now Christina has company of a sort. I understand from Dr. Herb Schaumburg, who is the first to describe the syndrome, that large numbers of patients are turning up everywhere now with severe sensory neuronopathies. The worst affected have body image disturbances like Christina. Most of them are health faddists, or are on a megavitamin craze, and have been taking enormous quantities of vitamin B_6 (pyridoxine). Thus there are now some hundreds of "disembodied" men and women—though most, unlike Christina, can hope to get better as soon as they stop poisoning themselves with pyridoxine.

4

The Man Who Fell Out of Bed

When I was a medical student many years ago, one of the nurses called me in considerable perplexity, and gave me this singular story on the phone: that they had a new patient—a young man—just admitted that morning. He had seemed very nice, very normal, all day—indeed, until a few minutes before, when he awoke from a snooze. He then seemed excited and strange—not himself in the least. He had somehow contrived to fall out of bed, and was now sitting on the floor, carrying on and vociferating, and refusing to go back to bed. Could I come, please, and sort out what was happening?

When I arrived, I found the patient lying on the floor by his bed and staring at one leg. His expression contained anger, alarm, bewilderment and amusement—bewilderment most of all, with a hint of consternation. I asked him if he would go back to bed, or if he needed help, but he seemed upset by these suggestions and shook his head. I squatted down beside him, and took the history on the floor. He had come in, that morning, for some tests, he said. He had no complaints, but the neurologists, feeling that he had a "lazy" left leg—that was the very word they had used—thought he should come in. He had felt fine all day and fallen asleep towards evening. When he woke up, he felt fine too, until he moved in the bed. Then he found, as he put it, "some-

one's leg" in the bed—*a severed human leg*, a horrible thing!
He was stunned, at first, with amazement and disgust—he
had never experienced, never imagined, such an incredible
thing. He felt the leg gingerly. It seemed perfectly formed,
but "peculiar" and cold. At this point he had a brainwave. He
now realized what had happened: *it was all a joke!* A rather
monstrous and improper, but a very original, joke! It was
New Year's Eve, and everyone was celebrating. Half the staff
were drunk; quips and crackers were flying; a carnival scene.
Obviously one of the nurses with a macabre sense of humor
had stolen into the Dissecting Room and nabbed a leg, and
then slipped it under his bedclothes as a joke while he was
still fast asleep. He was much relieved at the explanation;
but feeling that a joke was a joke, and that this one was a bit
much, he threw the damn thing out of the bed. But—and at
this point his conversational manner deserted him, and he
suddenly trembled and became ashen-pale—*when he threw
it out of bed, he somehow came after it—and now it was
attached to him.*

"Look at it!" he cried, with revulsion on his face. "Have you
ever seen such a creepy, horrible thing? I thought a cadaver
was just dead. But this is uncanny! And somehow—it's
ghastly—it seems stuck to me!" He seized it with both hands,
with extraordinary violence, and tried to tear it off his body,
and, failing, punched it in an access of rage.

"Easy!" I said. "Be calm! Take it easy! I wouldn't punch that
leg like that."

"And why not?" he asked, irritably, belligerently.

"Because it's *your* leg," I answered. "Don't you know your
own leg?"

He gazed at me with a look compounded of stupefaction,
incredulity, terror, and amusement, not unmixed with a jocu-

lar sort of suspicion. "Ah, Doc!" he said. "You're fooling me!
You're in cahoots with that nurse—you shouldn't kid patients
like this!"

"I'm not kidding," I said. "That's your own leg."

He saw from my face that I was perfectly serious—and a
look of utter terror came over him. "You say it's my leg, Doc?
Wouldn't you say that a man should know his own leg?"

"Absolutely," I answered. "He *should* know his own leg. I
can't imagine him *not* knowing his own leg. Maybe *you're* the
one who's been kidding all along?"

"I swear to God, cross my heart, I haven't. . . . A man *should*
know his own body, what's his and what's not—but this leg,
this *thing*"—another shudder of distaste—"doesn't feel right,
doesn't feel real—and it doesn't *look* part of me."

"What *does* it look like?" I asked in bewilderment, being, by
this time, as bewildered as he was.

"What does it look like?" He repeated my words slowly. "I'll
tell you what it looks like. *It looks like nothing on earth.* How
can a thing like that belong to me? I don't know *where* a thing
like that belongs. . . ." His voice trailed off. He looked terrified
and shocked.

"Listen," I said. "I don't think you're well. Please allow us to
return you to bed. But I want to ask you one final question. If
this—this thing—is *not* your left leg" (he had called it a "coun-
terfeit" at one point in our talk, and expressed his amazement
that someone had gone to such lengths to "manufacture" a
"facsimile") "then where *is* your own left leg?"

Once more he became pale—so pale that I thought he was
going to faint. "I don't know," he said. "I have no idea. It's disappeared. It's gone. It's nowhere to be found. . . ."

Since this account was first published (in *A Leg to Stand On*, 1984), I received a letter from the eminent neurologist Dr. Michael Kremer, who wrote:

> I was asked to see a puzzling patient on the cardiology ward. He had atrial fibrillation and had thrown off a large embolus giving him a left hemiplegia, and I was asked to see him because he constantly fell out of bed at night for which the cardiologists could find no reason.
>
> When I asked him what happened at night, he said quite openly that when he woke in the night he always found that there was a dead, cold, hairy leg in bed with him which he could not understand but could not tolerate and he, therefore, with his good arm and leg pushed it out of bed and naturally, of course, the rest of him followed.
>
> He was such an excellent example of this complete loss of awareness of his hemiplegic limb but, interestingly enough, I could not get him to tell me whether his own leg on that side was in bed with him because he was so caught up with the unpleasant foreign leg that was there.

5

Hands

Madeleine J. was admitted to St. Benedict's Hospital near New York City in 1980, her sixtieth year, a congenitally blind woman with cerebral palsy, who had been looked after by her family at home throughout her life. Given this history, and her pathetic condition—with spasticity and athetosis, i.e., involuntary movements of both hands, to which was added a failure of the eyes to develop—I expected to find her both retarded and regressed.

She was neither. Quite the contrary: she spoke freely, indeed eloquently (her speech, mercifully, was scarcely affected by spasticity), revealing herself to be a high-spirited woman of exceptional intelligence and literacy.

"You've read a tremendous amount," I said. "You must be really at home with Braille."

"No, I'm not," she said. "All my reading has been done for me—by talking books or other people. I can't read Braille, not a single word. I can't do *anything* with my hands—they are completely useless."

She held them up, derisively. "Useless godforsaken lumps of dough—they don't even feel part of me."

I found this very startling. The hands are not usually affected by cerebral palsy—at least, not essentially affected: they may be somewhat spastic, or weak, or deformed, but are generally of considerable use (unlike the legs, which may be completely paralyzed—in that variant called Little's disease, or cerebral diplegia).

Miss J.'s hands were *mildly* spastic and athetotic, but her sensory capacities—as I now rapidly determined—were completely intact: she immediately and correctly identified light touch, pain, temperature, passive movement of the fingers. There was no impairment of elementary sensation as such, but, in dramatic contrast, there was the profoundest impairment of perception. She could not recognize or identify anything whatever—I placed all sorts of objects in her hands, including one of my own hands. She could not identify—and she did not explore; there were no active "interrogatory" movements of her hands—they were, indeed, as inactive, as inert, as useless, as "lumps of dough."

This is very strange, I said to myself. How can one make sense of all this? There is no gross sensory "deficit." Her hands would seem to have the potential of being perfectly good hands—and yet they are not. Can it be that they are functionless—"useless"—because she had never used them? Had being "protected," "looked after," "babied" since birth prevented her from the normal exploratory use of the hands which all infants learn in the first months of life? Had she been carried about, had everything done for her, in a manner that had prevented her from developing a normal pair of hands? And if this was the case—it seemed far-fetched, but was the only hypothesis I could think of—could she now, in her sixtieth year, acquire what she should have acquired in the first weeks and months of life?

Was there any precedent? Had anything like this ever been described—or tried? I did not know, but I immediately thought of a possible parallel—what was described by Leont'ev and Zaporozhets in their book *Rehabilitation of Hand Function*.

The condition they were describing was quite different in origin: they described a similar "alienation" of the hands in some two hundred soldiers following massive injury and surgery—the injured hands felt "foreign," "lifeless," "useless," "stuck on," despite elementary neurological and sensory intactness. Leont'ev and Zaporozhets spoke of how the "gnostic systems" that allow "gnosis," or perceptive use of the hands, to take place could be "dissociated" in such cases as a consequence of injury, surgery and the weeks- or months-long hiatus in the use of the hands that followed. In Madeleine's case, although the phenomenon was identical—"uselessness," "lifelessness," "alienation"—it was lifelong. She did not need just to recover her hands, but to discover them—to acquire them, to achieve them—for the first time: not just to regain a dissociated gnostic system, but to construct a gnostic system she had never had in the first place. Was this possible?

The injured soldiers described by Leont'ev and Zaporozhets had normal hands before injury. All they had to do was to "remember" what had been "forgotten," or "dissociated," or "inactivated," through severe injury. Madeleine, in contrast, had no repertoire of memory, for she had never used her hands—and she felt she *had* no hands—or arms either. She had never fed herself, used the toilet by herself, or reached out to help herself, always leaving it for others to help her. She had behaved, for sixty years, as if she were a being without hands.

This then was the challenge that faced us: a patient with perfect elementary sensations in the hands, but apparently no power to integrate these sensations to the level of perceptions that were related to the world and to herself; no power to say, "I perceive, I recognize, I will, I act," so far as her "useless" hands went. But somehow or other (as Leont'ev and

Zaporozhets found with their patients), we had to get her to act and to use her hands actively, and, we hoped, in so doing to achieve integration: "The integration is in the action," as Roy Campbell said.

Madeleine was agreeable to all this, indeed fascinated, but puzzled and not hopeful. "How *can* I do anything with my hands," she asked, "when they are just lumps of putty?"

"In the beginning is the deed," Goethe writes. This may be so when we face moral or existential dilemmas, but not where movement and perception have their origin. Yet here too there is always something sudden: a first step (or a first word, as when Helen Keller said "water"), a first movement, a first perception, a first impulse—total, "out of the blue," where there was nothing, or nothing with sense before. "In the beginning is the impulse." Not a deed, not a reflex, but an "impulse," which is both more obvious and more mysterious than either. . . . We could not say to Madeleine, "Do it!" but we might hope for an impulse; we might hope for, we might solicit, we might even provoke one. . . .

I thought of the infant as it reached for the breast. "Leave Madeleine her food, as if by accident, slightly out of reach on occasion," I suggested to her nurses. "Don't starve her, don't tease her, but show less than your usual alacrity in feeding her." And one day it happened—what had never happened before: impatient, hungry, instead of waiting passively and patiently, she reached out an arm, groped, found a bagel, and took it to her mouth. This was the first use of her hands, her first manual act, in sixty years, and it marked her birth as a "motor individual" (Sherrington's term for the person who emerges through acts). It also marked her first manual perception, and thus her birth as a complete "perceptual individual." Her first perception, her first recognition, was of a bagel, or "bagelhood"—as Helen Keller's first recognition, first utterance, was of water ("waterhood").

After this first act, this first perception, progress was extremely rapid. As she had reached out to explore or touch a bagel, so now, in her new hunger, she reached out to explore or touch the whole world. Eating led the way—the feeling, the exploring, of different foods, containers, implements, etc. "Recognition" had somehow to be achieved by a curiously roundabout sort of inference or guesswork, for having been both blind and "handless" since birth, she was lacking in the simplest internal images (whereas Helen Keller at least had tactile images). Had she not been of exceptional intelligence and literacy, with an imagination filled and sustained, so to speak, by the images of others, images conveyed by language, by the *word*, she might have remained almost as helpless as a baby.

A bagel was recognized as round bread, with a hole in it; a fork as an elongated flat object with several sharp tines. But then this preliminary analysis gave way to an immediate intuition, and objects were instantly recognized as themselves, as immediately familiar in character and "physiognomy," were immediately recognized as unique, as "old friends." And this sort of recognition, not analytic but synthetic and immediate, went with a vivid delight and a sense that she was discovering a world full of enchantment, mystery and beauty.

The commonest objects delighted her—delighted her and stimulated a desire to reproduce them. She asked for clay and started to make models: her first model, her first sculpture, was of a shoehorn, and even this was somehow imbued with a peculiar power and humor, with flowing, powerful, chunky curves reminiscent of an early Henry Moore.

And then—and this was within a month of her first recognitions—her attention, her appreciation, moved from objects to people. There were limits, after all, to the interest and expressive possibilities of things, even when transfigured by a sort of innocent, ingenuous and often comical genius. Now she needed to explore the human face and figure, at rest and in motion. To be "felt" by Madeleine was a remarkable experience. Her hands, only such a little while ago inert, doughy, now seemed charged with a preternatural animation and sensibility. One was not merely being recognized, being scrutinized, in a way more intense and searching than any visual scrutiny, but being "tasted" and appreciated meditatively, imaginatively and aesthetically, by a born (a newborn) artist. They were, one felt, not just the hands of a blind woman exploring, but of a blind artist, a meditative and creative mind, just opened to the full sensuous and spiritual reality of the world. These explorations too pressed for representation and reproduction as an external reality.

She started to model heads and figures, and within a year was locally famous as the Blind Sculptress of St. Benedict's. Her sculptures tended to be half or three-quarters life size, with simple but recognizable features, and with a remarkably expressive energy. For me, for her, for all of us, this was a deeply moving, an amazing, almost a miraculous, experience. Who would have dreamed that basic powers of perception, normally acquired in the first months of life, but failing to be acquired at this time, could be acquired in one's sixtieth year? What wonderful possibilities of late learning, and learning for the handicapped, this opened up. And who could have dreamed that in this blind, palsied woman, hidden away, inactivated, overprotected all her life, there lay the germ of an astonishing artistic sensibility (unsuspected by her, as by others) that would germinate and blossom into a rare and beautiful reality, after remaining dormant, blighted, for sixty years?

POSTSCRIPT

The case of Madeleine J., however, as I was to find, was by no means unique. Within a year I had encountered another patient, Simon K., who also had cerebral palsy combined with profound impairment of vision. While Mr. K. had normal strength and sensation in his hands, he scarcely ever used them—and was extraordinarily inept at handling, exploring, or recognizing anything. Now we had been alerted by Madeleine J., we wondered whether he too might not have a similar "developmental agnosia"—and, as such, be "treatable" in the same way. And, indeed, we soon found that what had been achieved with Madeleine could be achieved with Simon as well. Within a year he had become very "handy" in all ways, and particularly enjoyed simple carpentry, shaping plywood and wooden blocks, and assembling them into simple wooden toys. He had no impulse to sculpt, to make reproductions—he was not a natural artist like Madeleine. But still, after a half century spent virtually without hands, he enjoyed their use in all sorts of ways.

This is the more remarkable, perhaps, because he is mildly retarded, an amiable simpleton, in contrast to the passionate and highly gifted Madeleine J. It might be said that she is extraordinary, a Helen Keller, a woman in a million—but nothing like this could possibly be said of simple Simon. And yet the essential achievement—the achievement of hands—proved wholly as possible for him as for her. It seems clear that intelligence, as such, plays no part in the matter—that the sole and essential thing is *use*.

Such cases of developmental agnosia may be rare, but one commonly sees cases of acquired agnosia, which illustrate the same fundamental principle of use. Thus I frequently see patients with a severe "glove-and-stocking" neuropathy, so-called, due to diabetes. If the neuropathy is sufficiently severe, patients go beyond feelings of numbness (the "glove-and-stocking" feeling), to a feeling of complete nothingness or de-realization. They may feel (as one patient put it) "like a basket case," with hands and feet completely "missing." Sometimes they feel their arms and legs end in stumps, with lumps of "dough" or "plaster" somehow "stuck on." Typically this feeling of de-realization, if it occurs, is absolutely sudden, and the return of reality, if it occurs, is equally sudden. There is, as it were, a critical (functional and ontological) threshold. It is crucial to get such patients to *use* their hands and feet—even, if necessary, to "trick" them into so doing. With this there is apt to occur a sudden re-realization—a sudden leap back into subjective reality and "life" . . . provided there is sufficient physiological potential (if the neuropathy is total, if the distal parts of the nerves are quite dead, no such re-realization is possible).

For patients with a severe but subtotal neuropathy, a modicum of use is literally vital, and makes all the difference between being a "basket case" and reasonably functional (with excessive use, there may be fatigue of the limited nerve function, and sudden derealization again).

It should be added that these subjective feelings have precise objective correlates: one finds "electrical silence," locally, in the muscles of the hands and feet; and, on the sensory side, a complete absence of any "evoked potentials," at every level up to the sensory cortex. As soon as the hands and feet are re-realized, with use, there is a complete reversal of the physiological picture.

A similar feeling of deadness and unrealness is described in Chapter 3, "The Disembodied Lady."

6

Phantoms

A "phantom," in the sense that neurologists use, is a persistent image or memory of part of the body, usually a limb, for months or years after its loss. Known in antiquity, phantoms were described and explored in great detail by the great American neurologist Silas Weir Mitchell, during and following the Civil War.

Weir Mitchell described several *sorts* of phantom—some strangely ghostlike and unreal (these were the ones he called "sensory ghosts"); some compellingly, even dangerously, lifelike and real; some intensely painful, others (most) quite painless; some photographically exact, like replicas or facsimiles of the lost limb, others grotesquely foreshortened or distorted . . . as well as "negative phantoms," or "phantoms of absence." He also indicated, clearly, that such "body image" disorders—the term was only introduced (by Henry Head) fifty years later—might be influenced by either central factors (stimulation or damage to the sensory cortex, especially that of the parietal lobes), or peripheral ones (the condition of the nerve stump, or neuromas; nerve damage, nerve block or nerve stimulation; disturbances in the spinal nerve roots or sensory tracts in the cord). I have been particularly interested, myself, in these peripheral determinants.

The following pieces, extremely short, almost anecdotal, were first published in the "Clinical Curio" section of the *British Medical Journal*.

PHANTOM FINGER

A sailor accidentally cut off his right index finger. For forty years afterwards he was plagued by an intrusive phantom of the finger rigidly extended, as it was when cut off. Whenever he moved his hand toward his face—for example, to eat or to scratch his nose—he was afraid that this phantom finger would poke his eye out. (He knew this to be impossible, but the feeling was irresistible.) He then developed severe sensory diabetic neuropathy and lost all sensation of even having any fingers. The phantom finger disappeared too.

It is well known that a central pathological disorder, such as a sensory stroke, can "cure" a phantom. How often does a peripheral pathological disorder have the same effect?

DISAPPEARING PHANTOM LIMBS

All amputees, and all who work with them, know that a phantom limb is essential if an artificial limb is to be used. Dr. Michael Kremer writes: "Its value to the amputee is enormous. I am quite certain that no amputee with an artificial lower limb can walk on it satisfactorily until the body-image, in other words the phantom, is incorporated into it."

Thus the disappearance of a phantom may be disastrous, and its recovery, its re-animation, a matter of urgency. This may be effected in all sorts of ways: Weir Mitchell describes how, with faradization of the brachial plexus, a phantom hand, missing for twenty-five years, was suddenly "resurrected." One such patient, under my care, describes how he must "wake up" his phantom in the mornings: first he flexes the thigh stump towards him, and then he slaps it sharply—"like a baby's bottom"—several times. On the fifth or sixth slap the phantom suddenly shoots forth, rekindled, *fulgurated*, by the peripheral stimulus. Only then can he put on his prosthesis and walk. What other odd methods (one wonders) are used by amputees?

POSITIONAL PHANTOMS

A patient, Charles D., was referred to us for stumbling, falls and vertigo—there had been unfounded suspicions of labyrinthine disorder. It was evident on closer questioning that what he experienced was not vertigo at all, but a flutter of ever-changing positional illusions—suddenly the floor seemed further, then suddenly nearer, it pitched, it jerked, it tilted—in his own words "like a ship in heavy seas." In consequence he found himself lurching and pitching, *unless he looked down at his feet.* Vision was necessary to show him the true position of his feet and the floor—feel had become grossly unstable and misleading—but sometimes even vision was overwhelmed by feel, so that the floor and his feet *looked* frightening and shifting.

We soon ascertained that he was suffering from the acute onset of *tabes*—and (in consequence of dorsal root involvement) from a sort of sensory delirium of rapidly fluctuating "proprioceptive illusions." Everyone is familiar with the classical end stage of *tabes*, in which there may be virtual proprioceptive "blindness" for the legs. Have readers encountered this intermediate stage—of positional phantoms or illusions due to an acute (and reversible) tabetic delirium?

The experience this patient recounts reminds me of a singular experience of my own, occurring with the *recovery* from a proprioceptive scotoma. This I described (in *A Leg to Stand On*) as follows:

I was infinitely unsteady, and had to gaze down. There and then I perceived the source of the commotion. The source

was my leg—or, rather, that thing, that featureless cylinder
of chalk which served as my leg—that chalky-white abstraction of a leg. Now the cylinder was a thousand feet long,
now a matter of two millimeters; now it was fat, now it was
thin; now it was tilted this way, now tilted that. It was constantly changing in size and shape, in position and angle, the
changes occurring four or five times a second. The extent of
transformation and change was immense—there could be a
thousandfold switch between successive "frames."

PHANTOMS—DEAD OR ALIVE?

There is often a certain confusion about phantoms—whether
they should occur, or not; whether they are pathological, or
not; whether they are "real," or not. The literature is confusing, but patients are not—and they clarify matters by describing different *sorts* of phantoms.

Thus a clear-headed man, with an above-the-knee amputation, described this to me:

There's this *thing*, this ghost-foot, which sometimes hurts
like hell—and the toes curl up, or go into spasm. This is
worst at night, or with the prosthesis off, or when I'm not
doing anything. It goes away, when I strap the prosthesis
on and walk. I still feel the leg then, vividly, but it's a *good*
phantom, different—it animates the prosthesis, and allows
me to walk.

With this patient, with all patients, is not *use* all-important,
in dispelling a "bad" (or passive, or pathological) phantom, if
it exists; and in keeping the "good" phantom—that is, the persisting personal limb-memory or limb-image—alive, active,
and well, as they need?

POSTSCRIPT

Many (but not all) patients with phantoms suffer "phantom pain," or pain in the phantom. Sometimes this has a bizarre quality, but often it is a rather "ordinary" pain, the persistence of a pain previously present in the limb, or the onset of a pain that might be expected were the limb actually present. I have—since the original publication of this book—received many fascinating letters from patients about this: one such patient speaks of the discomfort of an ingrown toenail, which had not been "taken care of" before amputation, persisting for years after the amputation; but also of an entirely different pain—an excruciating root pain or "sciatica" in the phantom—following an acute "slipped disc," and disappearing with removal of the disc and spinal fusion. Such problems, not at all uncommon, are in no sense "imaginary," and may indeed be investigated by neurophysiological means.

Thus Dr. Jonathan Cole, a former student of mine and now a spinal neurophysiologist, describes how in a woman with persistent phantom leg pain, anesthesia of the spinous ligament with Lignocaine caused the phantom to be anesthetized (indeed to disappear) briefly; but that electrical stimulation of the spinal roots produced a sharp tingling pain in the phantom quite different from the dull one which was usually present; whilst stimulation of the spinal cord higher up reduced the phantom pain (*personal communication*). Dr. Cole has also presented detailed electrophysiological studies of a patient with a sensory polyneuropathy of fourteen years' duration, very similar in many respects to Christina, the "Disembodied Lady." He tells the story of this patient in his book *Pride and a Daily Marathon*.

7

On the Level

It is nine years now since I met Mr. MacGregor, in the neurology clinic of St. Dunstan's, an old people's home where I once worked, but I remember him—I see him—as if it were yesterday.

"What's the problem?" I asked, as he tilted in.

"Problem? No problem—none that I know of. . . . But others keep telling me I lean to the side: 'You're like the Leaning Tower of Pisa,' they say. 'A bit more tilt, and you'll topple right over.'"

"But *you* don't feel any tilt?"

"I feel fine. I don't know what they mean. How *could* I be tilted without knowing I was?"

"It sounds a queer business," I agreed. "Let's have a look. I'd like to see you stand and take a little stroll—just from here to that wall and back. I want to see for myself, *and I want you to see too*. We'll take a videotape of you walking and play it right back."

"Suits me, Doc," he said, and, after a couple of lunges, stood up. What a fine old chap, I thought. Ninety-three—and he doesn't look a day past seventy. Alert, bright as a button. Good for a hundred. And strong as a coal-heaver, even if he does have Parkinson's disease. He was walking, now, confidently, swiftly, but canted over, improbably, a good twenty degrees, his center of gravity way off to the left, maintaining his balance by the narrowest possible margin.

"There!" he said with a pleased smile. "See! No problems—I walked straight as a die."

"Did you, indeed, Mr. MacGregor?" I asked. "I want you to judge for yourself."

I rewound the tape and played it back. He was profoundly shocked when he saw himself on the screen. His eyes bulged, his jaw dropped, and he muttered, "I'll be damned!" And then, "They're right, I *am* over to one side. I *see* it here clear enough, but I've no sense of it. I don't *feel* it."

"That's it," I said. "That's the heart of the problem."

We have five senses in which we glory and which we recognize and celebrate, senses that constitute the sensible world for us. But there are other senses—secret senses, sixth senses, if you will—equally vital, but unrecognized and unlauded. These senses, unconscious, automatic, had to be discovered. Historically, indeed, their discovery came late: what the Victorians vaguely called "muscle sense"—the awareness of the relative position of trunk and limbs, derived from receptors in the joints and tendons—was only really defined (and named "proprioception") in the 1890s. And the complex mechanisms and controls by which our bodies are properly aligned and balanced in space—these have only been defined in our own century, and still hold many mysteries. Perhaps it will only be in this space age, with the paradoxical license and hazards of gravity-free life, that we will truly appreciate our inner ears, our vestibules and all the other obscure receptors and reflexes that govern our body orientation. For normal man, in normal situations, they simply do not exist.

Yet their absence can be quite conspicuous. If there is defective (or distorted) sensation in our overlooked secret senses, what we then experience is profoundly strange, an almost incommunicable equivalent to being blind or being deaf. If proprioception is completely knocked out, the body becomes, so to speak, blind and deaf to itself—and (as the meaning of the Latin root *proprius* hints) ceases to "own" itself, to feel itself as itself (see "The Disembodied Lady").

The old man suddenly became intent, his brows knitted, his lips pursed. He stood motionless, in deep thought, presenting the picture that I love to see: a patient in the actual moment of discovery—half-appalled, half-amused—seeing for the first time exactly what is wrong and, in the same moment, exactly what there is to be done. This *is* the therapeutic moment.

"Let me think, let me think," he murmured, half to himself, drawing his shaggy white brows down over his eyes and emphasizing each point with his powerful, gnarled hands. "Let me think. You think with me—there must be an answer! I tilt to one side, and I can't tell it, right? There *should* be some feeling, a clear signal, but it's not there, right?" He paused. "I used to be a carpenter," he said, his face lighting up. "We would always use a spirit level to tell whether a surface was level or not, or whether it was tilted from the vertical or not. Is there a sort of spirit level in the brain?"

I nodded.

"Can it be knocked out by Parkinson's disease?"

I nodded again.

"Is *this* what has happened with me?"

I nodded a third time and said, "Yes. Yes. Yes."

In speaking of such a spirit level, Mr. MacGregor had hit on a fundamental analogy, a metaphor for an essential control system in the brain. Parts of the inner ear are indeed physically—literally—like levels; the labyrinth consists of semicircular canals containing liquid whose motion is continually monitored. But it was not these, as such, that were essentially at fault; rather, it was his ability to *use* his balance organs, in conjunction with the body's sense of itself and with its visual picture of the world. Mr. MacGregor's homely symbol applies not just to the labyrinth but also to the complex *integration* of the three secret senses: the labyrinthine, the proprioceptive, and the visual. It is this synthesis that is impaired in parkinsonism.

The most profound (and most practical) studies of such integrations—and of their singular *dis*integrations in parkinsonism—were made by James Purdon Martin and are to be found in his remarkable book *The Basal Ganglia and Posture* (originally published in 1967, but continually revised and expanded in the ensuing years; he was just completing a new edition when he died recently). Speaking of this integration, this integrator, in the brain, Purdon Martin writes, "There must be some centre or 'higher authority' in the brain . . . some 'controller' we may say. This controller or higher authority must be informed of the state of stability or instability of the body."

In the section on "tilting reactions" Purdon Martin emphasizes the threefold contribution to the maintenance of a stable and upright posture, and he notes how commonly its subtle balance is upset in parkinsonism—how, in particular, "it is usual for the labyrinthine element to be lost before the proprioceptive and the visual." This triple control system, he implies, is such that *one* sense, *one* control, can compensate for the others—not wholly (since the senses differ in their capabilities) but in part, at least, and to a useful degree. Visual reflexes and controls are perhaps the least important—normally. So long as our vestibular and proprioceptive systems are intact, we are perfectly stable with our eyes closed. We do not tilt or lean or fall over the moment we close our eyes. But the precariously balanced parkinsonian may do so. (One often sees parkinsonian patients sitting in the most grossly tilted positions, with no awareness that this is the case. But let a mirror be provided, so they can *see* their positions, and they instantly straighten up.)

Proprioception, to a considerable extent, can compensate for defects in the inner ears. Thus patients who have been surgically deprived of their labyrinths (as is sometimes done to relieve the intolerable, crippling vertigo of severe Ménière's disease), while at first unable to stand upright or take a single step, may learn to employ and to *enhance* their proprioception quite wonderfully; in particular, to use the sensors in the vast latissimus dorsi muscles of the back—the greatest, most mobile muscular expanse in the body—as an accessory and novel balance organ, a pair of vast, winglike proprioceptors. As the patients become practiced, as this becomes second nature, they are able to stand and walk—not perfectly, but with safety, assurance, and ease.

Purdon Martin was endlessly thoughtful and ingenious in designing a variety of mechanisms and methods that made it possible for even severely disabled parkinsonians to achieve an artificial normality of gait and posture—lines painted on the floor, counterweights in the belt, loudly ticking pacemakers—to set the cadence for walking. In this he always learned from his patients (to whom, indeed, his great book is dedicated). He was a deeply human pioneer, and in his medicine understanding and collaborating were central: patient and physician were coequals, on the same level, each learning from and helping the other and *between them* arriving at new insights and treatment. But he had not, to my knowledge, devised a prosthesis for the correction of impaired tilting and higher vestibular reflexes, the problem that afflicted Mr. MacGregor.

"So that's it, is it?" asked Mr. MacGregor. "I can't use the spirit level inside my head. I can't use my ears, but I *can* use my eyes." Quizzically, experimentally, he tilted his head to one side: "Things look the same now—the world doesn't tilt." Then he asked for a mirror, and I had a long one wheeled before him. "*Now* I see myself tilting," he said. "*Now* I can straighten up—maybe I could stay straight. . . . But I can't live among mirrors or carry one round with me."

He thought again deeply, frowning in concentration—then suddenly his face cleared, and lit up with a smile. "I've got it!" he exclaimed. "Yeah, Doc, I've got it! I don't need a mirror—I just need a level. I can't use the spirit levels *inside* my head, but why couldn't I use levels *outside* my head—levels I could *see*, I could use with my eyes?" He took off his glasses, fingering them thoughtfully, his smile slowly broadening.

"Here, for example, in the rim of my glasses . . . This could tell me, tell my eyes, if I was tilting. I'd keep an eye on it at first; it would be a real strain. But then it might become second nature, automatic. Okay, Doc, so what do you think?"

"I think it's a brilliant idea, Mr. MacGregor. Let's give it a try."

The principle was clear, the mechanics a bit tricky. We first experimented with a sort of pendulum, a weighted thread hung from the rims, but this was too close to the eyes, and scarcely seen at all. Then, with the help of our optometrist and workshop, we made a clip extending two nose-lengths forward from the bridge of the spectacles, with a miniature horizontal level fixed to each side. We fiddled with various designs, all tested and modified by Mr. MacGregor. In a couple of weeks we had completed a prototype, a pair of somewhat Heath Robinsonish spirit spectacles: "The world's first pair!" said Mr. MacGregor, in glee and triumph. He donned them. They looked a bit cumbersome and odd, but scarcely more so than the bulky hearing-aid spectacles that were coming in at the time. And now a strange sight was to be seen in our Home—Mr. MacGregor in the spirit spectacles he had invented and made, his gaze intensely fixed, like a steersman eyeing the binnacle of his ship. This worked, in a fashion—at least he stopped tilting: but it was a continuous, exhausting exercise. And then, over the ensuing weeks, it got easier and easier; keeping an eye on his "instruments" became unconscious, like keeping an eye on the instrument panel of one's car while being free to think, chat, and do other things.

Mr. MacGregor's spectacles became the rage of St. Dunstan's. We had several other patients with parkinsonism who also suffered from impairment of tilting reactions and postural reflexes—a problem not only hazardous but also notoriously resistant to treatment. Soon a second patient, then a third, were wearing Mr. MacGregor's spirit spectacles, and now, like him, could walk upright, on the level.

8

Eyes Right!

M rs. S., an intelligent woman in her sixties, has suffered a massive stroke, affecting the deeper and back portions of her right cerebral hemisphere. She has perfectly preserved intelligence—and humor.

She sometimes complains to the nurses that they have not put dessert or coffee on her tray. When they say, "But, Mrs. S., it is right there, on the left," she seems not to understand what they say, and does not look to the left. If her head is gently turned, so that the dessert comes into sight, in the preserved right half of her visual field, she says, "Oh, there is it—it wasn't there before." She has totally lost the idea of "left," with regard to both the world and her own body. Sometimes she complains that her portions are too small, but this is because she only eats from the right half of the plate—it does not occur to her that it has a left half as well. Sometimes, she will put on lipstick, and make up the right half of her face, leaving the left half completely neglected: it is almost impossible to treat these things, because her attention cannot be drawn to them ("hemi-inattention"—see Battersby 1956) and she has no conception that they are wrong. She knows it intellectually, and can understand, and laugh; but it is impossible for her to know it directly.

Knowing it intellectually, knowing it inferentially, she has worked out strategies for dealing with her imperception. She cannot look left, directly, she cannot turn left, so what she does is to turn right—and right through a circle. Thus she requested, and was given, a rotating wheelchair. And now if she cannot find something which she knows should be there, she swivels to the right, through a circle, until it comes into view. She finds this signally successful if she cannot find her coffee or dessert. If her portions seem too small, she will swivel to the right, keeping her eyes to the right, until the previously missed half now comes into view; she will eat this, or rather half of this, and feel less hungry than before. But if she is still hungry, or if she thinks on the matter, and realizes that she may have perceived only half of the missing half, she will make a second rotation till the remaining quarter comes into view, and, in turn, bisect this yet again. This usually suffices—after all, she has now eaten seven-eighths of the portion—but she may, if she is feeling particularly hungry or obsessive, make a third turn, and secure another sixteenth of her portion (leaving, of course, the remaining sixteenth, the left sixteenth, on her plate). "It's absurd," she says. "I feel like Zeno's arrow—I never get there. It may look funny, but under the circumstances what else can I do?"

It would seem far simpler for her to rotate the plate than rotate herself. She agrees, and has tried this—or at least tried to try it. But it is oddly difficult, it does not come naturally, whereas whizzing round in her chair does, because her looking, her attention, her spontaneous movements and impulses, are all now exclusively and instinctively to the right.

Especially distressing to her was the derision which greeted her when she appeared only half made-up, the left side of her face absurdly void of lipstick and rouge. "I look in the mirror," she said, "and do all I see." Would it be possible, we wondered, for her to have a "mirror" such that she would see the left side of her face on the right? That is, as someone else, facing her, would see her. We tried a video system, with camera and monitor facing her, and the results were startling, and bizarre. For now, using the video screen as a "mirror," she did see the left side of her face to her right, an experience confounding even to a normal person (as anyone knows who has tried to shave using a video screen), and doubly confounding, uncanny, for her, because the left side of her face and body, which she now saw, had no feeling, no existence, for her, in consequence of her stroke. "Take it away!" she cried, in distress and bewilderment, so we did not explore the matter further. This is a pity because, as R. L. Gregory also wonders, there might be much promise in such forms of video feedback for such patients with hemi-inattention and left hemi-field extinction. The matter is so physically, indeed metaphysically, confusing that only experiment can decide.

POSTSCRIPT

Computers and computer games (not available in 1976, when I saw Mrs. S.) may also be invaluable to patients with unilateral neglect in monitoring the "missing" half, or teaching them to do this themselves; I have recently (1986) made a short film of this.

I could not make reference, in the original edition of this book, to a very important book which came out almost simultaneously: *Principles of Behavioral Neurology*, edited by M. Marsel Mesulam. I cannot forbear quoting Mesulam's eloquent formulation of "neglect":

> When the neglect is severe, the patient may behave almost as if one half of the universe had abruptly ceased to exist in any meaningful form. . . . Patients with unilateral neglect behave not only as if nothing were actually happening in the left hemispace, but also as if nothing of any importance could be expected to occur there.

9

The President's Speech

What was going on? A roar of laughter from the aphasia ward, just as the President's speech was coming on, and they had all been so eager to hear the President speaking. . . .

There he was, the old Charmer, the Actor, with his practiced rhetoric, his histrionisms, his emotional appeal—and all the patients were convulsed with laughter. Well, not all: some looked bewildered, some looked outraged, one or two looked apprehensive, but most looked amused. The President was, as always, moving—but he was moving them, apparently, mainly to laughter. What could they be thinking? Were they failing to understand him? Or did they, perhaps, understand him all too well?

It was often said of these patients, who though intelligent had the severest receptive or global aphasia, rendering them incapable of understanding words as such, that they nonetheless understood most of what was said to them. Their friends, their relatives, the nurses who knew them well, could hardly believe, sometimes, that they *were* aphasic.

This was because, when addressed naturally, they grasped some or most of the meaning. And one does speak "naturally," naturally.

Thus, to demonstrate their aphasia, one had to go to extraordinary lengths, as a neurologist, to speak and behave un-naturally, to remove all the extraverbal cues: tone of voice, intonation, suggestive emphasis or inflection, as well as all visual cues (one's expressions, one's gestures, one's entire, largely unconscious, personal repertoire and posture). One had to remove all of this (which might involve total concealment of one's person, and total depersonalization of one's voice, even to using a computerized voice synthesizer) in order to reduce speech to pure words, speech totally devoid of what Frege called "tone-color" (*Klangenfarben*) or "evocation." With the most sensitive patients, it was only with such a grossly artificial, mechanical speech—somewhat like that of the computers in *Star Trek*—that one could be wholly sure of their aphasia.

Why all this? Because speech—natural speech—does *not* consist of words alone, nor (as Hughlings Jackson thought) "propositions" alone. It consists of *utterance*—an uttering-forth of one's whole meaning with one's whole being—the understanding of which involves infinitely more than mere word-recognition. And this was the clue to aphasiacs' understanding, even when they might be wholly uncomprehending of words as such. For though the words, the verbal constructions, *per se*, might convey nothing, spoken language is normally suffused with "tone," embedded in an expressiveness which transcends the verbal—and it is precisely this expressiveness, so deep, so various, so complex, so subtle, which is perfectly preserved in aphasia, though understanding of words be destroyed. Preserved—and often more: preternaturally enhanced.

This too becomes clear—often in the most striking, or comic, or dramatic way—to all those who work or live closely with aphasiacs: their families or friends or nurses or doctors. At first, perhaps, we see nothing much the matter; and then we see that there has been a great change, almost an inversion, in their understanding of speech. Something has gone, has been devastated, it is true—but something has come in its stead, has been immensely enhanced, so that—at least with emotionally laden utterance—the meaning may be fully grasped even when every word is missed. This, in our species *Homo loquens*, seems almost an inversion of the usual order of things: an inversion and perhaps a reversion too, to something more primitive and elemental. And this perhaps is why Hughlings Jackson compared aphasiacs to dogs (a comparison that might outrage both!)—though when he did this, he was chiefly thinking of their linguistic incompetences, rather than their remarkable and almost infallible sensitivity to "tone" and feeling. Henry Head, more sensitive in this regard, speaks of "feeling-tone" in his 1926 treatise on aphasia, and stresses how it is preserved, and often enhanced, in aphasiacs.*

Thus the feeling I sometimes have—which all of us who work closely with aphasiacs have—that one cannot lie to an aphasiac. He cannot grasp your words, and so cannot be deceived by them; but what he grasps he grasps with infallible precision, namely the *expression* that goes with the words, that total, spontaneous, involuntary expressiveness which can never be simulated or faked, as words alone can, all too easily.

We recognize this with dogs, and often use them for this

---

*"Feeling-tone" is a favorite term of Head's, which he uses in regard not only to aphasia but to the affective quality of sensation, as it may be altered by thalamic or peripheral disorders. Our impression, indeed, is that Head is continually half-unconsciously drawn towards the exploration of "feeling-tone"—towards, so to speak, a neurology of feeling-tone, in contrast or complementarity to a classical neurology of proposition and process. It is, incidentally, a common term in the United States, at least among blacks in the South: a common, earthy and indispensable term. "You see, there's such a thing as a feeling tone. . . . And if you don't have this, baby, you've had it" (cited by Studs Terkel as epigraph to his 1967 oral history *Division Street: America*).

purpose—to pick up falsehood, or malice, or equivocal inten-
tions, to tell us who can be trusted, who is integral, who
makes sense, when we—so susceptible to words—cannot trust
our own instincts.

And what dogs can do here, aphasiacs do too, and at a
human and immeasurably superior level. "One can lie with
the mouth," Nietzsche writes, "but with the accompanying
grimace one nevertheless tells the truth." To such a grimace,
to any falsity or impropriety in bodily appearance or posture,
aphasiacs are preternaturally sensitive. And if they cannot see
one—this is especially true of our blind aphasiacs—they have
an infallible ear for every vocal nuance, the tone, the rhythm,
the cadences, the music, the subtlest modulations, inflections,
intonations, which can give—or remove—verisimilitude to or
from a man's voice.

In this, then, lies their power of understanding—
understanding, without words, what is authentic or inau-
thentic. Thus it was the grimaces, the histrionisms, the false
gestures and, above all, the false tones and cadences of the
voice, which rang false for these wordless but immensely sensitive patients. It was to these (for them) most glaring, even
grotesque, incongruities and improprieties that my aphasic
patients responded, undeceived and undeceivable by words.

This is why they laughed at the President's speech.

If one cannot lie to an aphasiac, in view of his special sensitivity to expression and "tone," how is it, we might ask, with
patients—if there are such—who *lack* any sense of expression and "tone," while preserving, unchanged, their comprehension for words: patients of an exactly opposite kind? We
have a number of such patients, also on the aphasia ward,
although, technically, they do not have aphasia, but, instead,

a form of *agnosia*, in particular a so-called "tonal" agnosia. For such patients, typically, the expressive qualities of voices disappear—their tone, their timbre, their feeling, their entire character—while words and grammatical constructions are perfectly understood. Such tonal agnosias (or "aprosodias") are associated with disorders of the *right* temporal lobe of the brain, whereas the aphasias go with disorders of the *left* temporal lobe.

Among the patients with tonal agnosia on our aphasia ward who also listened to the President's speech was Emily D., with a glioma in her right temporal lobe. A former English teacher, and poetess of some repute, with an exceptional feeling for language and strong powers of analysis and expression, Emily D. was able to articulate the opposite situation—how the President's speech sounded to someone with tonal agnosia. Emily D. could no longer tell if a voice was angry, cheerful, sad—whatever. Since voices now lacked expression, she had to look at people's faces, their postures and movements when they talked, and found herself doing so with a care, an intensity, she had never shown before. But this, it so happened, was also limited, because she had a malignant glaucoma, and was rapidly losing her sight too.

What she then found she had to do was to pay extreme attention to exactness of words and word use, and to insist that those around her did just the same. She could less and less follow loose speech or slang—speech of an allusive or emotional kind—and more and more required of her interlocutors that they speak *prose*—"proper words in proper places." Prose, she found, might compensate, in some degree, for lack of perceived tone or feeling.

In this way she was able to preserve, even enhance, the use of "expressive" speech—in which the meaning was wholly given by the apt choice and reference of words—despite being more and more lost with "evocative" speech (where meaning is wholly given in the use and sense of tone).

Emily D. also listened, stony-faced, to the President's speech, bringing to it a strange mixture of enhanced and defective perceptions—precisely the opposite mixture to those of our aphasiacs. It did not move her—no speech now moved her—and all that was evocative, genuine or false completely passed her by. Deprived of emotional reaction, was she then (like the rest of us) transported or taken in? By no means. "He is not cogent," she said. "He does not speak good prose. His word-use is improper. Either he is brain damaged, or he has something to conceal." Thus the President's speech did not work for Emily D. either, due to her enhanced sense of formal language use, propriety as prose, any more than it worked for our aphasiacs, with their word-deafness but enhanced sense of tone.

Here then was the paradox of the President's speech. We normals—aided, doubtless, by our wish to be fooled, were indeed well and truly fooled ("*Populus vult decipi, ergo decipiatur*"). And so cunningly was deceptive word-use combined with deceptive tone, that only the brain damaged remained intact, undeceived.

PART TWO

Excesses

Introduction

D eficit," we have said, is neurology's favorite word— its only word, indeed, for any disturbance of function. Either the function (like a capacitor or fuse) is normal—or it is defective or faulty: what other possibility *is* there for a mechanistic neurology, which is essentially a system of capacities and connections?

What then of the opposite—an excess or superabundance of function? Neurology has no word for this—because it has no concept. A function, or functional system, works—or it does not: these are the only possibilities it allows. Thus a disease which is "ebullient" or "productive" in character challenges the basic mechanistic concepts of neurology, and this is doubtless one reason why such disorders—common, important, and intriguing as they are—have never received the attention they deserve. They receive it in psychiatry, where one speaks of excited and productive disorders—extravagances of fancy, of impulse . . . of mania. And they receive it in anatomy and pathology, where one speaks of hypertrophies, monstrosities—of teratoma. But physiology has no equivalent for this—no equivalent of monstrosities or manias. And this alone suggests that our basic concept or vision of the nervous system—as a sort of machine or computer—is radically inadequate, and needs to be supplemented by concepts more dynamic, more alive.

This radical inadequacy may not be apparent when we consider only loss—the privation of functions we considered in Part I. But it becomes immediately obvious if we consider their excesses—not amnesia, but hypermnesia; not agnosia, but hypergnosia; and all the other "hypers" we can imagine.

Classical, "Jacksonian" neurology never considers such disorders of excess—that is, primary superabundances or burgeonings of functions (as opposed to so-called "releases"). Hughlings Jackson himself, it is true, did speak of "hyper-physiological" and "super-positive" states. But here, we might say, he is letting himself go, being playful, or simply just being faithful to his clinical experience, though at odds with his own mechanical concepts of function (such contradictions were characteristic of his genius, the chasm between his naturalism and his rigid formalism).

We have to come almost to the present day to find a neurologist who even *considers* an excess. Thus Luria's two clinical biographies are nicely balanced: *The Man with a Shattered World* is about loss, *The Mind of a Mnemonist* about excess. I find the latter by far the more interesting and original of the two, for it is, in effect, an exploration of imagination and memory (and no such exploration is possible to classical neurology).

In *Awakenings* there was an internal balance, so to speak, between the terrible privations seen before L-Dopa—akinesia, aboulia, adynamia, anergia, etc.—and the almost equally terrible excesses after L-Dopa—hyperkinesia, hyperboulia, hyperdynamia, etc.

And in this we see the emergence of a new sort of term, of terms and concepts other than those of function—impulse, will, dynamism, energy—terms essentially kinetic and dynamic (whereas those of classical neurology are essentially static).

And, in the mind of the Mnemonist, we see dynamisms of a much higher order at work—the thrust of an ever-burgeoning and almost uncontrollable association and imagery, a monstrous growth of thinking, a sort of teratoma of the mind, which the Mnemonist himself calls an "It."

But the word "It," or automatism, is also too mechanical. "Burgeoning" conveys better the disquietingly alive quality of the process. We see in the Mnemonist—or in my own overenergized, galvanized patients on L-Dopa—a sort of animation gone extravagant, monstrous, or mad—not merely an excess, but an organic proliferation, a generation; not just an imbalance, a disorder of function, but a disorder of generation.

We might imagine, from a case of amnesia or agnosia, that there is merely a function or competence impaired—but we see from patients with hypermnesias and hypergnosias that mnesis and gnosis are inherently active and generative at all times; inherently, and—potentially—monstrously as well. Thus, we are forced to move from a neurology of function to a neurology of action, of life. This crucial step is forced upon us by the diseases of excess—and without it we cannot begin to explore the "life of the mind." Traditional neurology, by its mechanicalness, its emphasis on deficits, conceals from us the actual life which is instinct in all cerebral functions—at least higher functions such as those of imagination, memory and perception. It conceals from us the very life of the mind. It is with these living (and often highly personal) dispositions of brain and mind—especially in a state of enhanced, and thus illuminated, activity—that we shall be concerned now.

Enhancement allows the possibilities not only of a healthy fullness and exuberance but of a rather ominous extravagance, aberration, monstrosity—the sort of "too-muchness" which continually loomed in *Awakenings*, as patients, overexcited, tended to disintegration and uncontrol; an overpowering by impulse, image and will; possession (or dispossession) by a physiology gone wild.

This danger is built into the very nature of growth and life. Growth can become over-growth, life "hyper-life." All the "hyper" states can become monstrous, perverse aberrations, "para" states: hyperkinesia tends towards parakinesia—abnormal movements, chorea, tics; hypergnosia readily becomes paragnosia—perversions, apparitions, of the morbidly heightened senses; the ardors of "hyper" states can become violent passions.

The paradox of an illness which can present as wellness—as a wonderful feeling of health and well-being, and only later reveal its malignant potentials—is one of the chimaeras, tricks and ironies of nature. It is one which has fascinated a number of artists, especially those who equate art with sickness: thus it is a theme—at once Dionysiac, Venerean and Faustian—which persistently recurs in Thomas Mann—from the febrile tuberculous highs of *The Magic Mountain*, to the spirochetal inspirations in *Dr. Faustus* and the aphrodisiac malignancy in his last tale, *The Black Swan*.

I have always been intrigued by such ironies and have written of them before. In *Migraine* I spoke of the high which may precede, or constitute the start of, attacks—and quoted George Eliot's remark that feeling "dangerously well" was often, for her, the sign or harbinger of an attack. "Dangerously well"—what an irony is this: it expresses precisely the doubleness, the paradox, of feeling "too well."

For "wellness," naturally, is no cause for complaint—people relish it, they enjoy it, they are at the furthest pole from complaint. People complain of feeling ill—not well. Unless, as George Eliot does, they have some intimation of "wrongness," or danger, either through knowledge or association, or the very excess of excess. Thus, though a patient will scarcely complain of being "very well," they may become suspicious if they feel "too well."

This was a central, and (so to speak) cruel, theme in *Awakenings*, that patients profoundly ill, with the profoundest deficits, for many decades, might find themselves, as by a miracle, suddenly well, only to move from there into the hazards, the tribulations, of excess, functions stimulated far beyond "allowable" limits. Some patients realized this, had premonitions—but some did not. Thus Rose R., in the first flush and joy of restored health, said, "It's fabulous, it's gorgeous!"; but as things accelerated towards uncontrol, she said, "Things can't last. Something awful is coming." And similarly, with more or less insight, in most of the others—as with Leonard L., as *he* passed from repletion to excess. As I wrote in *Awakenings*, "his abundance of health and energy—of 'grace,' as he called it—became *too* abundant, and started to assume an extravagant form. His sense of harmony and ease and effortless control was replaced by a sense of *too-muchness* . . . a great surplus, a great *pressure* of . . . [every kind]," which threatened to disintegrate him, to burst him asunder.

This is the simultaneous gift and affliction, the delight, the anguish, conferred by excess. And it is felt, by insightful patients, as questionable and paradoxical: "I have too much energy," one Tourette patient said. "Everything is too bright, too powerful, too much. It is a feverish energy, a morbid brilliance."

"Dangerous wellness," "morbid brilliance," a deceptive euphoria with abysses beneath—*this* is the trap promised and threatened by excess, whether it be set by nature, in the form of some intoxicating disorder, or by ourselves in the form of some excitant addiction.

The human dilemmas, in such situations, are of an extraordinary kind: for patients are here faced with disease as seduction, something remote from, and far more equivocal than, the traditional theme of illness as suffering or affliction. And nobody, absolutely nobody, is exempt from such bizarrenesses, such indignities. In disorders of excess there may be a sort of collusion in which the self is more and more aligned and identified with its sickness, so that finally it seems to lose all independent existence and be nothing but a product of sickness. This *fear* is expressed by Witty Ticcy Ray when he says, "I consist of tics—there is nothing else," or when he envisages a mind-growth—a "Tourettoma"—which might engulf him. For him, with his strong ego and relatively mild Tourette's syndrome, there was not, in reality, any such danger. But for patients with weak or undeveloped egos, coupled with overwhelmingly strong disease, there is a very real risk of such "possession" or "dispossession." I do no more than touch on this in "The Possessed."

10

Witty Ticcy Ray

In 1885 Gilles de la Tourette, a pupil of Charcot, described the astonishing syndrome which now bears his name. "Tourette's syndrome," as it was immediately dubbed, is characterized by an excess of nervous energy, and a great production and extravagance of strange motions and notions: tics, jerks, mannerisms, grimaces, noises, curses, involuntary imitations and compulsions of all sorts, with an odd elfin humor and a tendency to antic and outlandish kinds of play. In its "highest" forms, Tourette's syndrome involves every aspect of the affective, the instinctual, and the imaginative life; in its "lower" and perhaps commoner forms, there may be little more than abnormal movements and impulsivity, though even here there is an element of strangeness. It was well recognized and extensively reported in the closing years of the last century, for these were years of a spacious neurology which did not hesitate to conjoin the organic and the psychic. It was clear to Tourette and his peers that this syndrome was a sort of possession by primitive impulses and urges, but also that it was a possession with an organic basis—a very definite (if undiscovered) neurological disorder.

In the years that immediately followed the publication of Tourette's original papers many hundreds of cases of this syndrome were described—no two cases ever being quite the same. It became clear that there were forms which were mild

and benign, and others of quite terrible grotesqueness and
violence. Equally, it was clear that some people could "take"
Tourette's and accommodate it within a commodious personality, even gaining advantage from the swiftness of thought
and association and invention which went with it, while others might indeed be "possessed" and scarcely able to achieve
real identity amid the tremendous pressure and chaos of
Tourettic impulses. There was always, as Luria remarked of
his mnemonist, a fight between an "It" and an "I."

Charcot and his pupils (who included Freud and Babinski
as well as Tourette) were among the last of their profession
with a combined vision of body and soul, "It" and "I," neurology and psychiatry. By the turn of the century, a split had
occurred, into a soulless neurology and a bodiless psychology,
and with this any understanding of Tourette's disappeared. In
fact, Tourette's syndrome itself seemed to have disappeared,
and was scarcely at all reported in the first half of this century.
Some physicians, indeed, regarded it as "mythical," a product
of Tourette's colorful imagination; most had never heard of
it. It was as forgotten as the great sleepy-sickness epidemic of
the 1920s.

The forgetting of sleepy-sickness (*encephalitis lethargica*)
and the forgetting of Tourette's have much in common. Both
disorders were extraordinary, and strange beyond belief—at
least, the beliefs of a contracted medicine. They could not
be accommodated in the conventional frameworks of medicine, and therefore they were forgotten and mysteriously "disappeared." But there is a much more intimate connection,
which was hinted at in the 1920s, in the hyperkinetic or frenzied forms which the sleepy-sickness sometimes took: these
patients tended, at the beginning of their illness, to show a
mounting excitement of mind and body, violent movements,

tics, compulsions of all kinds. Some time afterwards, they were overtaken by an opposite fate, an all-enveloping trance-like "sleep"—in which I found them forty years later.

In 1969, I gave these sleepy-sickness or postencephalitic patients L-Dopa, a precursor of the transmitter dopamine, which was greatly lowered in their brains. They were transformed by it. First they were "awakened" from stupor to health; then they were driven towards the other pole—of tics and frenzy. This was my first experience of Tourette-like syndromes: wild excitements, violent impulses, often combined with a weird, antic humor. I started to speak of "Tourettism," although I had never seen a patient with Tourette's.

Early in 1971, the *Washington Post*, which had taken an interest in the "awakening" of my postencephalitic patients, asked me how they were getting on. I replied, "They are tic-cing," which prompted them to publish an article on "Tics." After the publication of this article, I received countless letters, the majority of which I passed on to my colleagues. But there was one patient I did consent to see—Ray.

The day after I saw Ray, it seemed to me that I noticed three Touretters in the street in downtown New York. I was confounded, for Tourette's syndrome was said to be excessively rare. It had an incidence, I had read, of one in a million, yet I had apparently seen three examples in an hour. I was thrown into a turmoil of bewilderment and wonder: was it possible that I had been overlooking this all the time, either not seeing such patients or vaguely dismissing them as "nervous," "cracked," "twitchy"? Was it possible that everyone had been overlooking them? Was it possible that Tourette's was not a rarity, but rather common—a thousand times more common, say, than previously supposed? The next day, without specially looking, I saw another two in the street. At this

point I conceived a whimsical fantasy or private joke: suppose
(I said to myself) that Tourette's is very common yet fails to
be recognized—but once recognized is easily and constantly
seen.* Suppose one such Touretter recognizes another, and
these two a third, and these three a fourth, until, by incre-
menting recognition, a whole band of them is found: brothers
and sisters in pathology, a new species in our midst, joined
together by mutual recognition and concern. Could there not
come together, by such spontaneous aggregation, a whole
association of New Yorkers with Tourette's?

Three years later, in 1974, I found that my fantasy had
become a reality: that there had indeed come into being a
Tourette's Syndrome Association. It had fifty members then:
now, seven years later, it has a few thousand. This astounding increase must be ascribed to the efforts of the TSA itself,
even though it consists only of patients, their relatives, and
physicians. The association has been endlessly resourceful in
its attempts to make known (or, in the best sense, "publicize")
the Touretter's plight. It has aroused responsible interest
and concern in place of the repugnance, or dismissal, which
had so often been the Touretter's lot, and it has encouraged
research of all kinds, from the physiological to the sociological: research into the biochemistry of the Tourettic brain, on
genetic and other factors which may co-determine Tourette's,
on the abnormally rapid and indiscriminate associations and
reactions which characterize it. Instinctual and behavioral

---

*A very similar situation happened with muscular dystrophy, which
was never seen until Duchenne described it in the 1850s. By 1860, after
his original description, many hundreds of cases had been recognized
and described, so much so that Charcot said, "How come that a disease
so common, so widespread, and so recognisable at a glance—a disease
which has doubtless always existed—how come that it is only recognised
now? Why did we need M. Duchenne to open our eyes?"

structures of a developmentally and even phylogenetically primitive kind have been revealed. There has been research on the body language and grammar and linguistic structure of tics; there have been unexpected insights into the nature of cursing and joking (which are also characteristic of some other neurological disorders); and, not least, there have been studies of the interaction of Touretters with their family and others, and of the strange mishaps which may attend these relationships. The TSA's remarkably successful endeavors are an integral part of the history of Tourette's and, as such, unprecedented: never before have patients led the way to understanding, become the active and enterprising agents of their own comprehension and cure.

What has emerged in these last ten years—largely under the aegis and stimulus of the TSA—is a clear confirmation of Gilles de la Tourette's intuition that this syndrome indeed has an organic neurological basis. The "It" in Tourette's, like the "It" in parkinsonism and chorea, reflects what Pavlov called "the blind force of the subcortex," a disturbance of those primitive parts of the brain which govern "go" and "drive." In parkinsonism, which affects motion but not action as such, the disturbance lies in the midbrain and its connections. In chorea, which is a chaos of fragmentary quasi-actions, the disorder lies in higher levels of the basal ganglia. In Tourette's, where there is excitement of the emotions and the passions, a disorder of the primal, instinctual bases of behavior, the disturbance seems to lie in the very highest parts of the "old brain": the thalamus, hypothalamus, limbic system, and amygdala, where the basic affective and instinctual determinants of personality are lodged. Thus Tourette's, pathologically no less than clinically, constitutes a sort of "missing link" between body and mind, and lies, so to speak, between chorea and mania. As in the rare, hyperkinetic forms of *encephalitis lethargica* and in all postencephalitic patients overexcited by L-Dopa, patients with Tourette's syndrome, or "Tourettism" from any other cause (strokes, cerebral tumors, intoxications or infections), seem to have an excess of excitor transmitters in the brain, especially the transmitter dopamine. And as lethargic parkinsonian patients need more dopamine to arouse them, as my postencephalitic patients were "awakened" by the dopamine-precursor L-Dopa, so frenetic and Tourettic patients must have their dopamine lowered by a dopamine antagonist, such as the drug haloperidol (Haldol).

On the other hand, there is not just a surfeit of dopamine in the Touretter's brain, as there is not just a deficiency of it in the parkinsonian brain. There are also much subtler and more widespread changes, as one would expect in a disorder which may alter personality: there are countless subtle paths of abnormality which differ from patient to patient, and from day to day in any one patient. Haldol can be an answer to Tourette's, but neither it nor any other drug can be *the* answer, any more than L-Dopa is *the* answer to parkinsonism. Complementary to any purely medicinal or medical approach there must also be an "existential" approach—in particular, a sensitive understanding of action, art, and play as being in essence healthy and free, and thus antagonistic to crude drives and impulsions, to "the blind force of the subcortex" from which these patients suffer. The motionless parkinsonian can sing and dance, and when he does so is completely free from his parkinsonism; and when the galvanized Touretter sings, plays or acts, he in turn is completely liberated from his Tourette's. Here the "I" vanquishes and reigns over the "It."

Between 1973 and his death in 1977, I enjoyed the privilege of corresponding with the great neuropsychologist A. R. Luria, and often sent him observations, and tapes, on Tourette's. In one of his last letters, he wrote to me: "This is truly of a tremendous importance. Any understanding of such a syndrome must vastly broaden our understanding of human nature in general. . . . I know of no other syndrome of comparable interest."

When I first saw Ray he was 24 years old, and almost incapacitated by multiple tics of extreme violence coming in volleys every few seconds. He had been subject to these since the age of four and severely stigmatized by the attention they aroused, though his high intelligence, his wit, his strength of character and sense of reality enabled him to pass successfully through school and college, and to be valued and loved by a few friends and his wife. Since leaving college, however, he had been fired from a dozen jobs—always because of tics, never for incompetence—was continually in crises of one sort and another, usually caused by his impatience, his pugnacity, and his coarse, brilliant "chutzpah." And he had found his marriage threatened by involuntary cries of "Fuck!" "Shit!," and so on, which would burst from him at times of sexual excitement. He was (like many Touretters) remarkably musical, and could scarcely have survived—emotionally or economically—had he not been a weekend jazz drummer of real virtuosity, famous for his sudden and wild extemporizations, which would arise from a tic or a compulsive hitting of a drum and would instantly be made the nucleus of a wild and wonderful improvisation, so that the "sudden intruder" would be turned to brilliant advantage. His Tourette's was also of advantage in various games, especially ping-pong, at which he excelled, partly in consequence of his abnormal quickness of reflex and reaction, but especially, again, because of "improvisations," "very sudden, nervous, *frivolous* shots" (in his own words), which were so unexpected and startling as to be virtually unanswerable. The only time he was free from tics was in post-coital quiescence or in sleep; or when he swam or sang or worked, evenly and rhythmically, and found "a kinetic melody," a play, which was tension-free, tic-free and free.

Under an ebullient, eruptive, clownish surface, he was a deeply serious man—and a man in despair. He had never heard of the TSA (which, indeed, scarcely existed at the time), nor had he heard of Haldol. He had diagnosed himself as having Tourette's after reading the article on "Tics" in the *Washington Post*. When I confirmed the diagnosis, and spoke of using Haldol, he was excited but cautious. I made a test of Haldol by injection, and he proved extraordinarily sensitive to it, becoming virtually tic-free for a period of two hours after I had administered no more than one-eighth of a milligram. After this auspicious trial, I started him on Haldol, prescribing a dose of a quarter of a milligram three times a day.

He came back the following week, with a black eye and a broken nose, and said: "So much for your fucking Haldol." Even this minute dose, he said, had thrown him off balance, interfered with his speed, his timing, his preternaturally quick reflexes. Like many Touretters, he was attracted to spinning things, and to revolving doors in particular, which he would dodge in and out of like lightning, but he had lost this knack on the Haldol, had mistimed his movements, and had been bashed on the nose. Further, many of his tics, far from disappearing, had simply become slow, and enormously extended: he might get "transfixed in mid-tic," as he put it, and find himself in almost catatonic postures (Ferenczi once called catatonia the opposite of tics—and suggested these be called "cataclonia"). He presented a picture, even on this minute dose, of marked

parkinsonism, dystonia, catatonia and psychomotor "block,"
a reaction which seemed inauspicious in the extreme, sug-
gesting not insensitivity but such over-sensitivity, such pathological sensitivity, that perhaps he could only be thrown from
one extreme to another—from acceleration and Tourettism to
catatonia and parkinsonism, with no possibility of any happy
medium.

He was understandably discouraged by this experience,
and this thought, and also by another thought which he now
expressed. "Suppose you *could* take away the tics," he said.
"What would be left? I consist of tics—there is nothing else."
He seemed, at least jokingly, to have little sense of his identity
except as a ticqueur: he called himself "the ticcer of President's Broadway" and spoke of himself, in the third person, as
"witty ticcy Ray," adding that he was so prone to "ticcy witti-
cisms and witty ticcicisms" that he scarcely knew whether it
was a gift or a curse. He said he could not imagine life without
Tourette's, nor was he sure he would care for it.

I was strongly reminded, at this point, of what I had encountered in some of my postencephalitic patients, who were inordinately sensitive to L-Dopa. I had nevertheless observed in
their case that such extreme physiological sensitivities and
instabilities might be transcended if it were possible for the
patient to lead a rich and full life—that the "existential" balance or poise of such a life might overcome a severe physiological imbalance. Feeling that Ray also had such possibilities
in him, that, despite his own words, he was not incorrigibly
centered on his own disease in an exhibitionistic or narcissis-
tic way, I suggested that we meet weekly for a period of three
months. During this time we would try to imagine life without
Tourette's; we would explore (if only in thought and feeling)
how much life could offer, could offer *him*, without the per-

verse attractions and attentions of Tourette's; we would examine the role and economic importance of Tourette's to him,
and how he might get on without these. We would explore all
this for three months—and then make another trial of Haldol.

There followed three months of deep and patient exploration, in which (often against much resistance and spite and
lack of faith in self and life) all sorts of healthy and human
potentials came to light: potentials which had somehow survived twenty years of severe Tourette's and "Touretty" life,
hidden in the deepest and strongest core of the personality.
This deep exploration was exciting and encouraging in itself
and gave us at least a limited hope.

What in fact happened exceeded all our expectations and
showed itself to be no mere flash in the pan, but an enduring
and permanent transformation of reactivity. For when I again
tried Ray on Haldol, in the same minute dose as before, he
now found himself tic-free, but without significant ill-effects—
and he has remained this way for the past nine years.

The effects of Haldol, here, were "miraculous"—but only
became so when a miracle was allowed. Its initial effects
were close to catastrophic—partly, no doubt, on a physiological basis; but also because any "cure" or relinquishing
of Tourette's at this time would have been premature and
economically impossible. Having had Tourette's since the age
of four, Ray had no experience of any normal life—he was
heavily dependent on his exotic disease and, not unnaturally,
employed and exploited it in various ways. He had not been
ready to give up his Tourette's and (I cannot help thinking)
might never have been ready without those three months of
intense preparation, of tremendously hard and concentrated
deep analysis and thought.

The past nine years, on the whole, have been happy ones

for Ray—a liberation beyond any possible expectation. After twenty years of being confined by Tourette's, and compelled to this and that by its crude physiology, he enjoys a spaciousness and freedom he would never have thought possible (or at most, during our analysis, only theoretically possible). His marriage is tender and stable, and he is now a father as well; he has many good friends, who love and value him as a person and not simply as an accomplished Tourettic clown; he plays an important part in his local community; and he holds a responsible position at work. Yet problems remain, problems perhaps inseparable from having Tourette's—and Haldol.

During his working hours, and working week, Ray remains "sober, solid, square" on Haldol—this is how he describes his "Haldol self." He is slow and deliberate in his movements and judgments, with none of the impatience, the impetuosity, he showed before Haldol but, equally, none of the wild improvisations and inspirations. Even his dreams are different in quality: "straight wish-fulfilment," he says, "with none of the elaborations, the extravaganzas, of Tourette's." He is less sharp, less quick in repartee, no longer bubbling with witty tics or ticcy wit. He no longer enjoys or excels at ping-pong or other games; he no longer feels "that urgent killer instinct, the instinct to win, to beat the other man"; he is less competitive, then, and also less playful; and he has lost the impulse or the knack of sudden "frivolous" moves which take everyone by surprise. He has lost his obscenities, his coarse chutzpah, his spunk. He has come to feel, increasingly, that something is missing.

Most important, and disabling, because this was vital for him as a means of both support and self-expression, he found that on Haldol he was musically "dull"—average, competent, but lacking energy, enthusiasm, extravagance, and joy. He no longer had tics or compulsive hitting of the drums—but he no longer had wild and creative surges.

As this pattern became clear to him, and after discussing it with me, Ray made a momentous decision: he would take Haldol "dutifully" throughout the working week but would take himself off it and "let fly" at weekends. This he has done for the past three years. So now there are two Rays—on and off Haldol. There is the sober citizen, the calm deliberator, from Monday to Friday; and there is "witty ticcy Ray," frivolous, frenetic, inspired, at weekends. It is a strange situation, as Ray is the first to admit:

> Having Tourette's [he says] is wild, like being drunk all the while. Being on Haldol is dull, makes one square and sober, and neither state is really free. . . . You "normals," who have the right transmitters in the right places at the right times in your brains, have all feelings, all styles, available all the time—gravity, levity, whatever is appropriate. We Touretters don't: we are forced into levity by our Tourette's and forced into gravity when we take Haldol. *You* are free, you have a natural balance; we must make the best of an artificial balance.

Ray does make the best of it and has a full life, despite Tourette's, despite Haldol, despite the "unfreedom" and the "artifice," despite being deprived of that birthright of natural freedom which most of us enjoy. But he has been taught by his sickness and, in a way, he has transcended it. He would say, with Nietzsche, "I have traversed many kinds of health, and keep traversing them. . . . And as for sickness: are we not almost tempted to ask whether we could get along without it? Only great pain is the ultimate liberator of the spirit."

Paradoxically, Ray—deprived of natural, animal physiological health—has found a new health, a new freedom, through the vicissitudes he is subject to. He has achieved what Nietzsche liked to call "the Great Health"—rare humor, valor, and resilience of spirit—despite being, or because he is, afflicted with Tourette's.

11

Cupid's Disease

A bright woman of ninety, Natasha K., recently came to our clinic. Soon after her eighty-eighth birthday, she said, she noticed "a change." What sort of change? we queried.

"Delightful!" she exclaimed. "I thoroughly enjoyed it. I felt more energetic, more alive—I felt young once again. I took an interest in the young men. I started to feel, you might say, 'frisky'—yes, frisky."

"This was a problem?"

"No, not at first. I felt well, *extremely* well—why should I think anything was the matter?"

"And then?"

"My friends started to worry. First they said, 'You look radiant—a new lease on life!' but then they started to think it was not quite . . . appropriate. 'You were always so shy,' they said, 'and now you're a flirt. You giggle, you tell jokes—at your age, is that right?'"

"And how did *you* feel?"

"I was taken aback. I'd been carried along, and it didn't occur to me to question what was happening. But then I did. I said to myself, 'You're 89, Natasha, this has been going on for a year. You were always so temperate in feeling—and now this extravagance! You are an old woman, nearing the end. What could justify such a sudden euphoria?' And as soon as I thought of euphoria, things took on a new complexion. . . .

'You're sick, my dear,' I said to myself. 'You're feeling *too* well, you have to be ill!' "

"Ill? Emotionally? Mentally ill?"

"No, not emotionally—physically ill. It was something in my body, my brain, that was making me high. And then I thought—goddam it, it's Cupid's Disease!"

"Cupid's Disease?" I echoed, blankly. I had never heard of the term before.

"Yes, Cupid's Disease—syphilis, you know. I was in a brothel in Salonika, nearly seventy years ago. I caught syphilis—lots of the girls had it—we called it Cupid's Disease. My husband saved me, took me out, had it treated. That was years before penicillin, of course. Could it have caught up with me after all these years?"

There may be an immense latent period between the primary infection and the advent of neurosyphilis, especially if the primary infection has been suppressed, not eradicated. I had one patient, treated with Salvarsan by Ehrlich himself, who developed tabes dorsalis—one form of neurosyphilis—more than fifty years later.

But I had never heard of an interval of *seventy* years—nor of a self-diagnosis of cerebral syphilis mooted so calmly and clearly.

"That's an amazing suggestion," I replied after some thought. "It would never have occurred to me—but perhaps you are right."

She was right; the spinal fluid was positive, she did have neurosyphilis, it *was* indeed the spirochetes stimulating her ancient cerebral cortex. Now the question of treatment arose. But here another dilemma presented itself, propounded, with typical acuity, by Mrs. K. herself. "I don't know that I *want* it treated," she said. "I know it's an illness, but it's made me

feel *well*. I've enjoyed it, I still enjoy it, I won't deny it. It's
made me feel livelier, friskier, than I have in twenty years.
It's been fun. But I know when a good thing goes too far,
and stops being good. I've had thoughts, I've had impulses,
I won't tell you, which are . . . well, embarrassing and silly.
It was like being a little tiddly, a little tipsy, at first, but if it
goes any further . . ." She mimed a drooling, spastic dement.
"I guessed I had Cupid's, that's why I came to you. I don't
want it to get worse, that would be awful; but I don't want
it cured—that would be just as bad. I wasn't fully alive until
the wrigglies got me. *Do you think you could keep it just as
it is?*"

We thought for a while, and our course, mercifully, was
clear. We have given her penicillin, which has killed the spirochetes, but can do nothing to reverse the cerebral changes, the
disinhibitions, they have caused.

And now Mrs. K. has it both ways, enjoying a mild disinhibition, a release of thought and impulse, without any threat to
her self-control or of further damage to her cortex. She hopes
to live, thus reanimated, rejuvenated, to a hundred. "Funny
thing," she says. "You've got to give it to Cupid."

POSTSCRIPT

Very recently (January 1985) I have seen some of these same
dilemmas and ironies in relation to another patient (Miguel O.),
admitted to the state hospital with a diagnosis of "mania," but
soon realized to be suffering from the excited stage of neurosyphilis. A simple man, he had been a farmhand in Puerto Rico,
and with some speech and hearing impediment, he could not
express himself too well in words, but expressed himself, exhibited his situation, simply and clearly, in drawings.

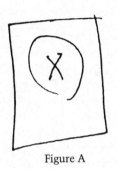

Figure A

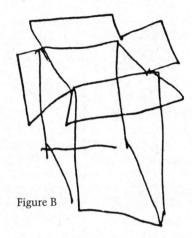

Figure B

Excited elaboration ("an open carton")

The first time I saw him he was quite excited, and when I asked him to copy a simple figure (Figure A) he produced, with great brio, a three-dimensional elaboration (Figure B)—or so I took it to be, until he explained that it was "an open carton," and then tried to draw some fruit in it. Impulsively inspired by his excited imagination, he had ignored the circle and cross, but retained and made concrete the idea of "enclosure." An open carton, a carton full of oranges—was that not more exciting, more alive, more real, than my dull figure?

A few days later I saw him again, very energized, very active, thoughts and feelings flying everywhere, high as a kite. I asked him again to draw the same figure. And now, impulsively, without pausing for a moment, he transformed the original to a sort of trapezoid, a lozenge, and then attached to this a string—and a boy (Figure C, next page). "Boy flying kite, kites flying!" he exclaimed excitedly.

I saw him for the third time a few days after this, and

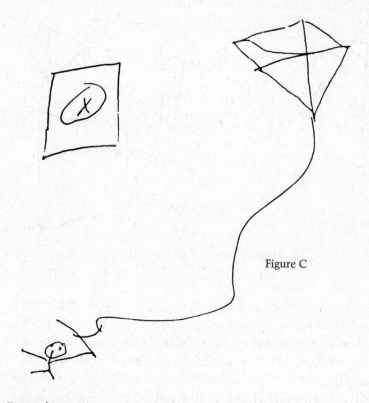

Figure C

Drugged, treated . . .
Imagination and animation gone

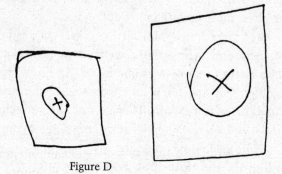

Figure D

found him rather down, rather parkinsonian (he had been given Haldol to quiet him, while awaiting final tests on the spinal fluid). Again I asked him to draw the figure, and this time he copied it dully, correctly, and a little smaller than the original (the "micrographia" of Haldol), and with none of the elaborations, the animation, the imagination, of the others (Figure D). "I don't 'see' things anymore," he said. "It looked so real, it looked so *alive* before. Will everything seem dead when I am treated?"

The drawings of patients with parkinsonism, as they are "awakened" by L-Dopa, form an instructive analogy. Asked to draw a tree, the parkinsonian tends to draw a small, meager thing, stunted, impoverished, a bare winter-tree with no foliage at all. As he "warms up," "comes to," is animated by L-Dopa, so the tree acquires vigor, life, imagination—and foliage. If he becomes too excited, high, on L-Dopa, the tree may acquire a fantastic ornateness and exuberance, exploding with a florescence of new branches and foliage with little arabesques, curlicues, and what-not, until finally its original form is completely lost beneath this enormous, this baroque, elaboration. Such drawings are also rather characteristic of Tourette's—the original form, the original thought, lost in a jungle of embellishment—and in the so-called "speed art" of amphetaminism. First the imagination is awakened, then excited, frenzied, to endlessness and excess.

What a paradox, what a cruelty, what an irony, there is here—that inner life and imagination may lie dull and dormant unless released, awakened, by an intoxication or disease!

Precisely this paradox lay at the heart of *Awakenings*; it is responsible too for the seduction of Tourette's and, no doubt, for the peculiar uncertainty which may attach to a drug like cocaine (which is known, like L-Dopa, or Tourette's, to raise the brain's dopamine). Thus Freud's startling comment about cocaine, that the sense of well-being and euphoria it induces "in no way differs from the normal euphoria of the healthy person. . . . In other words, you are simply normal, and it is soon hard to believe that you are under the influence of any drug."

The same paradoxical valuation may attach to electrical stimulations of the brain: there are epilepsies which are exciting and addictive, and may be self-induced, repeatedly, by those who are prone to them (as rats, with implanted cerebral electrodes, compulsively stimulate the "pleasure centers" of their own brain); but there are other epilepsies which bring peace and genuine well-being. A wellness can be genuine even if caused by an illness. And such a paradoxical wellness may even confer a lasting benefit, as with Mrs. O'C. and her strange convulsive "reminiscence" (Chapter 15).

We are in strange waters here, where all the usual considerations may be reversed—where illness may be wellness, and normality illness, where excitement may be either bondage or release, and where reality may lie in ebriety, not sobriety. It is the very realm of Cupid and Dionysus.

12

A Matter of Identity

W hat'll it be today?" he says, rubbing his hands. "Half a
     pound of Virginia, a nice piece of Nova?"

(Evidently he saw me as a customer—he would often pick
up the phone on the ward and say, "Thompson's Delicates-
sen.")

"Oh, Mr. Thompson!" I exclaim. "And who do you think
I am?"

"Good heavens, the light's bad—I took you for a customer.
As if it isn't my old friend Tom Pitkins. . . . Me and Tom" (he
whispers in an aside to the nurse) "was always going to the
races together."

"Mr. Thompson, you are mistaken again."

"So I am," he rejoins, not put out for a moment. "Why
would you be wearing a white coat if you were Tom? You're
Hymie, the kosher butcher next door. No bloodstains on your
coat though. Business bad today? You'll look like a slaughter-
house by the end of the week!"

Feeling a bit swept away myself in this whirlpool of identities, I finger the stethoscope dangling from my neck.

"A stethoscope!" he exploded. "And you pretending to be
Hymie! You mechanics are all starting to fancy yourselves to
be doctors, what with your white coats and stethoscopes—as
if you need a stethoscope to listen to a car! So, you're my old
friend Manners from the Mobil station up the block, come in
to get your bologna-and-rye. . . ."

William Thompson rubbed his hands again, in his salesman-grocer's gesture, and looked for the counter. Not finding it, he looked at me strangely again.

"Where am I?" he said, with a sudden scared look. "I thought I was in my shop, doctor. My mind must have wandered. . . . You'll be wanting my shirt off, to sound me as usual?"

"No, not the usual. I'm *not* your usual doctor."

"Indeed you're not. I could see that straightaway! You're not my usual chest-thumping doctor. And, by God, you've a beard! You look like Sigmund Freud—have I gone bonkers, round the bend?"

"No, Mr. Thompson. Not round the bend. Just a little trouble with your memory—difficulties remembering and recognizing people."

"My memory has been playing me some tricks," he admitted. "Sometimes I make mistakes—I take somebody for somebody else. . . . What'll it be now—Nova or Virginia?"

So it would happen, with variations, every time—with improvisations, always prompt, often funny, sometimes brilliant, and ultimately tragic. Mr. Thompson would identify me—misidentify, pseudo-identify me—as a dozen different people in the course of five minutes. He would whirl, fluently, from one guess, one hypothesis, one belief, to the next, without any appearance of uncertainty at any point—he never knew who I was, or what and where *he* was, an ex-grocer with severe Korsakov's in a neurological institution.

He remembered nothing for more than a few seconds. He was continually disoriented. Abysses of amnesia continually opened beneath him, but he would bridge them, nimbly, by fluent confabulations and fictions of all kinds. For him they were not fictions, but how he suddenly saw, or interpreted, the world. Its radical flux and incoherence could not be tolerated, acknowledged, for an instant—there was, instead, this strange, delirious, quasi-coherence, as Mr. Thompson, with his ceaseless, unconscious, quick-fire inventions, continually improvised a world around him—an Arabian Nights world, a phantasmagoria, a dream of ever-changing people, figures, situations—continual, kaleidoscopic mutations and transformations. For Mr. Thompson, however, it was not a tissue of ever-changing, evanescent fancies and illusion, but a wholly normal, stable, and factual world. So far as *he* was concerned, there was nothing the matter.

On one occasion, Mr. Thompson went for a trip, identifying himself at the front desk as "the Revd. William Thompson," ordering a taxi, and taking off for the day. The taxi driver, whom we later spoke to, said he had never had so fascinating a passenger, for Mr. Thompson told him one story after another, amazing personal stories full of fantastic adventures. "He seemed to have been everywhere, done everything, met everyone. I could hardly believe so much was possible in a single life," he said. "It is not exactly a single life," we answered. "It is all very curious—a matter of identity."*

Jimmie G., another Korsakov's patient, whom I have already described at length, had long since *cooled down* from his acute Korsakov's syndrome and seemed to have settled into a state of permanent lostness (or perhaps a permanent now-seeming dream or reminiscence of the past). But Mr. Thompson, only just out of hospital—his Korsakov's had exploded

*A very similar story is related by Luria in *The Neuropsychology of Memory*, in which the spellbound cabdriver only realized that his exotic passenger was ill when he gave him, for a fare, a temperature chart he was holding. Only then did he realize that this Scheherazade, this spinner of a thousand and one tales, was one of "those strange patients" at the Neurological Institute.

just three weeks before, when he developed a high fever, raved, and ceased to recognize all his family—was still on the boil, was still in an almost frenzied confabulatory delirium (of the sort sometimes called "Korsakov's psychosis," though it is not really a psychosis at all), continually creating a world and self to replace what was continually being forgotten and lost. Such a frenzy may call forth quite brilliant powers of invention and fancy—a veritable confabulatory genius—for such a patient *must literally make himself (and his world) up every moment*. We have, each of us, a life-story, an inner narrative whose continuity, whose sense, *is* our lives. It might be said that each of us constructs and lives a "narrative," and that this narrative *is* us, our identities.

If we wish to know about a man, we ask "what is his story—his real, inmost story?"—for each of us *is* a biography, a story. Each of us *is* a singular narrative, which is constructed, continually, unconsciously, by, through, and in us—through our perceptions, our feelings, our thoughts, our actions; and, not least, our discourse, our spoken narrations. Biologically, physiologically, we are not so different from each other; historically, as narratives, we are each of us unique.

To be ourselves we must *have* ourselves—possess, if need be re-possess, our life-stories. We must "recollect" ourselves, recollect the inner drama, the narrative, of ourselves. A man *needs* such a narrative, a continuous inner narrative, to maintain his identity, his self.

This narrative need, perhaps, is the clue to Mr. Thompson's desperate tale-telling, his verbosity. Deprived of continuity, of a quiet, continuous, inner narrative, he is driven to a sort of narrational frenzy—hence his ceaseless tales, his confabulations, his mythomania. Unable to maintain a genuine narrative or continuity, unable to maintain a genuine inner world, he is driven to the proliferation of pseudo-narratives, in a pseudo-continuity, pseudo-worlds peopled by pseudo-people, phantoms.

What is it *like* for Mr. Thompson? Superficially, he comes over as an ebullient comic. People say, "He's a riot." And there *is* much that is farcical in such a situation, which might form the basis of a comic novel.* It *is* comic, but not just comic—it is terrible as well. For here is a man who, in some sense, is desperate, in a frenzy. The world keeps disappearing, losing meaning, vanishing—and he must seek meaning, *make* meaning, in a desperate way, continually inventing, throwing bridges of meaning over abysses of meaninglessness, the chaos that yawns continually beneath him.

But does Mr. Thompson himself know this, feel this? After finding him "a riot," "a laugh," "loads of fun," people are disquieted, even terrified, by something in him. "He never stops," they say. "He's like a man in a race, a man trying to catch something which always eludes him." And indeed, he can never stop running, for the breach in memory, in existence, in meaning, is never healed, but has to be bridged, to be "patched," every second. And the bridges, the patches, for all their brilliance, fail to work, because they *are* confabulations, fictions, which cannot do service for reality, while also failing to correspond with reality. Does Mr. Thompson feel *this*? Or, again, what *is* his "feeling of reality"? Is he in a torment all the

---

*Indeed, such a novel has been written. Shortly after "The Lost Mariner" was published, a young writer named David Gilman sent me the manuscript of his book *Croppy Boy*, the story of an amnesiac like Mr. Thompson, who enjoys the wild and unbridled license of creating identities, new selves, as he whims, and as he must—an astonishing imagination of an amnesiac genius, told with positively Joycean richness and gusto. I do not know whether it has been published; I am very sure it should be. I could not help wondering whether Mr. Gilman had actually met (and studied) a "Thompson"—as I have often wondered whether Borges's "Funes," so uncannily similar to Luria's Mnemonist, may have been based on a personal encounter with such a mnemonist.

while, the torment of a man lost in unreality, struggling to rescue himself but sinking himself by ceaseless inventions, illusions, themselves quite unreal? It is certain that he is not at ease; there is a tense, taut look on his face all the while, as of a man under ceaseless inner pressure—and occasionally, not too often, or masked if present, a look of open, naked, pathetic bewilderment. What saves Mr. Thompson in a sense, and in another sense damns him, *is* the forced or defensive superficiality of his life: the way in which it is, in effect, reduced to a surface—brilliant, shimmering, iridescent, ever-changing, but for all that a surface, a mass of illusions, a delirium without depth.

And with this, no feeling *that* he has lost feeling (for the feeling he has lost), no feeling *that* he has lost the depth, that unfathomable, mysterious, myriad-leveled depth which somehow defines identity or reality. This strikes everyone who has been in contact with him for any time: that under his fluency, even his frenzy, is a strange loss of feeling—that feeling or judgment which distinguishes between "real" and "unreal," "true" and "untrue" (one cannot speak of "lies" here, only of "non-truth"), important and trivial, relevant and irrelevant. What comes out torrentially in his ceaseless confabulation has, finally, a peculiar quality of indifference—as if it didn't really matter what he said, or what anyone else did or said; as if nothing really mattered anymore.

A striking example of this was presented one afternoon, when William Thompson, jabbering away of all sorts of people who were improvised on the spot, said, "And there goes my younger brother, Bob, past the window," in the same excited but even and indifferent tone as the rest of his monologue. I was dumbfounded when, a minute later, a man peeked round the door, and said, "I'm Bob, I'm his younger brother; I think he saw me passing by the window." Nothing in William's tone or manner—nothing in his exuberant but unvarying and indifferent style of monologue—had prepared me for the possibility of . . . reality. William spoke of his brother, who *was* real, in precisely the same tone or lack of tone in which he spoke of the unreal—and now, suddenly, out of the phantoms, a real figure appeared! Further, he did not treat his younger brother as "real"—did not display any real emotion, was not in the least oriented or delivered from his delirium—but, on the contrary, instantly treated his brother as unreal, effacing him, losing him, in a further whirl of delirium. (This was utterly different from the rare but profoundly moving times when Jimmie, the lost mariner, met *his* brother, and while with him was unlost.) This was intensely disconcerting to poor Bob, who said, "I'm Bob, not Rob, not Dob," to no avail whatever. In the midst of confabulations—perhaps some strand of memory, of remembered kinship or identity was still holding, or came back for an instant—William spoke of his *elder* brother, George, using his invariable present indicative tense.

"But George died nineteen years ago!" said Bob, aghast.

"Aye, George is always the joker!" William quipped, apparently ignoring or indifferent to Bob's comment, and went on blathering of George in his excited, dead way, insensitive to truth, to reality, to propriety, to everything—insensitive, too, to the manifest distress of the living brother before him.

It was this which convinced me, above everything, that there was some ultimate and total loss of inner reality, of feeling and meaning, of soul, in William—and led me to ask the Sisters, as I had asked them of Jimmie G., "Do you think William *has* a soul? Or has he been pithed, scooped-out, de-souled, by disease?"

This time, however, they looked worried by my question,

as if something of the sort were already in their minds. They could not say, "Judge for yourself. See Willie in Chapel," because his wisecracking, his confabulations continued even there. There is an utter pathos, a sad *sense* of lostness, with Jimmie G., which one does not feel, or feel directly, with the effervescent Mr. Thompson. Jimmie has *moods*, and a sort of brooding (or, at least, yearning) sadness, a depth, a soul, which does not seem to be present in Mr. Thompson. Doubtless, as the Sisters said, he had a soul, an immortal soul, in the theological sense; could be seen, and loved, as an individual by the Almighty; but, they agreed, something very disquieting had happened to him, to his spirit, his character, in the ordinary, human sense.

It is *because* Jimmie is "lost" that he *can* be redeemed or found, at least for a while, in the mode of a genuine emotional relation. Jimmie is in despair, a quiet despair (to use or adapt Kierkegaard's term), and therefore he has the possibility of salvation, of touching base, the ground of reality, the feeling and meaning he has lost, but still recognizes, still yearns for.

But for William—with his brilliant, brassy surface, the unending joke which he substitutes for the world (which if it covers over a desperation, is a desperation he does not feel); for William, with his manifest indifference to relation and reality caught in an unending verbosity, there may be nothing "redeeming" at all—his confabulations, his apparitions, his frantic search for meanings, being the ultimate barrier to any meaning.

Paradoxically, then, William's great gift for confabulation— which has been called out to leap continually over the ever-opening abyss of amnesia—William's great gift is also his damnation. If only he could be *quiet*, one feels, for an instant; if only he could stop the ceaseless chatter and jabber; if only he could relinquish the deceiving surface of illusions—then (ah then!) reality might seep in; something genuine, something deep, something true, something felt, could enter his soul.

For it is not memory which is the final, "existential" casualty here (although his memory *is* wholly devastated); it is not memory only which has been so altered in him, but some ultimate capacity for feeling which is gone; and this is the sense in which he is "de-souled."

Luria speaks of such indifference as "equalization"—and sometimes seems to see it as the ultimate pathology, the final destroyer of any world, any self. It exerted, I think, a horrified fascination on him, as well as constituting an ultimate therapeutic challenge. He was drawn back to this theme again and again—sometimes in relation to Korsakov's and memory, as in *The Neuropsychology of Memory*, more often in relation to frontal lobe syndromes, especially in *Human Brain and Psychological Processes*, which contains several full-length case histories of such patients, fully comparable in their terrible coherence and impact to "the man with a shattered world"— comparable, and, in a way, more terrible still, because they depict patients who do not realize that anything has befallen them, patients who have lost their own reality without knowing it, patients who may not suffer but be the most godforsaken of all. Zazetsky (in *The Man with a Shattered World*) is constantly described as a *fighter*, always (even passionately) conscious of his state, and always fighting "with the tenacity of the damned" to recover the use of his damaged brain. But William (like Luria's frontal lobe patients—see the next chapter) is so damned he does not know he is damned, for it is not just a faculty, or some faculties, which are damaged but the very citadel, the self, the soul itself. William is "lost," in this sense, far more than Jimmie; for all his brio, one never feels, or rarely feels, that there is a *person* remaining, whereas in Jimmie there is plainly a real, moral being, even if disconnected most of the time. In Jimmie, at least, reconnection is *possible*—the therapeutic challenge can be summed up as "Only connect."

Our efforts to "reconnect" William all fail—even increase his confabulatory pressure. But when we abdicate our efforts and let him be, he sometimes wanders out into the quiet and undemanding garden which surrounds the Home, and there, in its quietness, he recovers his own quiet. The presence of others, other people, excite and rattle him, force him into an endless, frenzied social chatter, a veritable delirium of identity-making and -seeking; the presence of plants, a quiet garden, the non-human order, making no social or human demands upon him, allow this identity-delirium to relax, to subside; and by their quiet, non-human self-sufficiency and completeness allow him a rare quietness and self-sufficiency of his own, by offering (beneath or beyond all merely human identities and relations) a deep wordless communion with nature itself, and with this the restored sense of being in the world, being real.

13

Yes, Father-Sister

Mrs. B., a former research chemist, had presented with a rapid personality change, becoming "funny" (facetious, given to wisecracks and puns), impulsive—and "superficial." ("You feel she doesn't care about you," one of her friends said. "She no longer seems to care about anything at all.") At first it was thought that she might be hypomanic, but she turned out to have a cerebral tumor. At craniotomy there was found not a meningioma, as had been hoped, but a huge carcinoma involving the orbitofrontal aspects of both frontal lobes.

When I saw her, she seemed high-spirited, volatile—"a riot" (the nurses called her)—full of quips and cracks, often clever and funny.

"Yes, Father," she said to me on one occasion.

"Yes, Sister," on another.

"Yes, Doctor," on a third.

She seemed to use the terms interchangeably.

"What *am* I?" I asked, stung, after a while.

"I see your face, your beard," she said, "I think of an Archimandrite Priest. I see your white uniform—I think of the Sisters. I see your stethoscope—I think of a doctor."

"You don't look at *all* of me?"

"No, I don't look at all of you."

"You realize the difference between a father, a sister, a doctor?"

"I *know* the difference, but it means nothing to me. Father, sister, doctor—what's the big deal?"

Thereafter, teasingly, she would say, "Yes, father-sister. Yes, sister-doctor," and other combinations.

Testing left-right discrimination was oddly difficult, because she said "left" or "right" indifferently (though there was not, in reaction, any confusion of the two, as when there is a lateralizing defect of perception or attention). When I drew her attention to this, she said, "Left/right. Right/left. Why the fuss? What's the difference?"

"*Is* there a difference?" I asked.

"Of course," she said, with a chemist's precision. "You could call them *enantiomorphs* of each other. But they mean nothing to *me*. They're no different for *me*. Hands . . . Doctors . . . Sisters," she added, seeing my puzzlement. "Don't you understand? They mean nothing—nothing to me. *Nothing means anything* . . . at least to me."

"And . . . this meaning nothing . . ." I hesitated, afraid to go on. "This meaninglessness . . . does *this* bother you? Does *this* mean anything to you?"

"Nothing at all," she said promptly, with a bright smile, in the tone of one who makes a joke, wins an argument, wins at poker.

Was this denial? Was this a brave show? Was this the "cover" of some unbearable emotion? Her face bore no deeper expression whatever. Her world had been voided of feeling and meaning. Nothing any longer felt "real" (or "unreal"). Everything was now "equivalent" or "equal"—the whole world reduced to a facetious insignificance.

I found this somewhat shocking—her friends and family did too—but she herself, though not without insight, was uncaring, indifferent, even with a sort of funny-dreadful nonchalance or levity.

Mrs. B., though acute and intelligent, was somehow not

present—"de-souled"—as a person. I was reminded of William Thompson (and also of Dr. P.). This is the effect of the "equalization" described by Luria which we saw in the preceding chapter and will also see in the next.

POSTSCRIPT

The sort of facetious indifference and "equalization" shown by this patient is not uncommon—German neurologists call it *Witzelsucht* ("joking disease"), and it was recognized as a fundamental form of nervous "dissolution" by Hughlings Jackson a century ago. It is not uncommon, whereas insight is—and the latter, perhaps mercifully, is lost as the "dissolution" progresses. I see many cases a year with similar phenomenology but the most varied etiologies. Occasionally I am not sure, at first, if the patient is just "being funny," clowning around, or schizophrenic. Thus, almost at random, I find the following in my notes on a patient with cerebral multiple sclerosis, whom I saw (but whose case I could not follow up) in 1981:

> She speaks very quickly, impulsively, and (it seems) indifferently . . . so that the important and the trivial, the true and the false, the serious and the joking, are poured out in a rapid, unselective, half-confabulatory stream. . . . She may contradict herself completely within a few seconds . . . will say she loves music, she doesn't, she has a broken hip, she hasn't. . . .

I concluded my observation on a note of uncertainty:

> How much is cryptamnesia-confabulation, how much frontal lobe indifference-equalization, how much some strange schizophrenic disintegration and shattering-flattening?

Of all forms of "schizophrenia" the "silly-happy," the so-called "hebephrenic," most resembles the organic amnestic and frontal lobe syndromes. They are the most malignant, and the least imaginable—and no one returns from such states to tell us what they were like.

In all these states—"funny" and often ingenious as they appear—the world is taken apart, undermined, reduced to anarchy and chaos. There ceases to be any "center" to the mind, though its formal intellectual powers may be perfectly preserved. The end point of such states is an unfathomable "silliness," an abyss of superficiality, in which all is ungrounded and afloat and comes apart. Luria once spoke of the mind as reduced, in such states, to "mere Brownian movement." I share the sort of horror he clearly felt about them (though this incites, rather than impedes, their accurate description). They make me think, first, of Borges's "Funes," and his remark, "My memory, Sir, is like a garbage-heap," and finally, of the *Dunciad*, the vision of a world reduced to Pure Silliness—Silliness as being the End of the World:

> *Thy hand, great Anarch, lets the curtain fall;*
> *And Universal Darkness buries All.*

14

The Possessed

With Witty Ticcy Ray, I described a relatively mild form of Tourette's syndrome, but hinted that there were severer forms "of quite terrible grotesqueness and violence." I suggested that some people could accommodate Tourette's within a commodious personality, while others "might indeed be 'possessed,' and scarcely able to achieve real identity amid the tremendous pressure and chaos of Tourettic impulses."

Tourette himself, and many of the older clinicians, used to recognize a malignant form of Tourette's, which might disintegrate the personality and lead to a bizarre, phantasmagoric, pantomimic and often impersonatory form of "psychosis" or frenzy. This form of Tourette's—"super-Tourette's"—is quite rare, perhaps fifty times rarer than ordinary Tourette's syndrome, and it may be qualitatively different, as well as far more intense than any of the ordinary forms of the disorder. This "Tourette psychosis," this singular identity-frenzy, is quite different from ordinary psychosis, because of its underlying and unique physiology and phenomenology. Nonetheless it has affinities, on the one hand, to the frenzied motor psychoses sometimes induced by L-Dopa and, on the other, to the confabulatory frenzies of Korsakov's psychosis. And like these it can almost overwhelm the person.

The day after I saw Ray, my first Touretter, my eyes and mind opened, as I mentioned earlier, when, in the streets of

New York, I saw no less than three Touretters—all as characteristic as Ray, though more florid. It was a day of visions
for the neurological eye. In swift vignettes I witnessed what
it might mean to have Tourette's syndrome of ultimate sever-
ity, not only tics and convulsions of movement but tics and
convulsions of perception, imagination, the passions—of the
entire personality.

Ray himself had shown what might happen in the street.
But it is not enough to be told. You must see for yourself.
And a doctor's clinic or ward is not always the best place for
observing disease—at least not for observing a disorder which,
if organic in origin, is expressed in impulse, imitation, imper-
sonation, reaction, interaction, raised to an extreme and
almost incredible degree. The clinic, the laboratory, the ward
are all designed to restrain and focus behavior, if not indeed
to exclude it altogether. They are for a systematic and scientific neurology, reduced to fixed tests and tasks, not for an
open, naturalistic neurology. For this one must see the patient
unselfconscious, unobserved, in the real world, wholly given
over to the spur and play of every impulse, and one must oneself, the observer, be unobserved. What could be better, for
this purpose, than a street in New York—an anonymous public
street in a vast city—where the subject of extravagant, impulsive disorders can enjoy and exhibit to the full the monstrous
liberty, or slavery, of their condition.

"Street neurology," indeed, has respectable antecedents.
James Parkinson, as inveterate a walker of the streets of London as Charles Dickens was to be forty years later, delineated
the disease that bears his name not in his office but in the
teeming streets of London. Parkinsonism, indeed, cannot be
fully seen, comprehended, in the clinic; it requires an open,
complexly interactional space for the full revelation of its

peculiar character (beautifully shown in Jonathan Miller's film *Ivan*). Parkinsonism has to be seen, to be fully comprehended, in the world, and if this is true of parkinsonism, how much truer must it be of Tourette's. Indeed, an extraordinary description from within of an imitative and antic *ticqueur* in the streets of Paris is given in "*Les confidences d'un ticqueur*" which prefaces Meige and Feindel's great 1901 book *Tics*, and a vignette of a manneristic *ticqueur*, also in the streets of Paris, is provided by the poet Rilke in *The Notebook of Malte Laurids Brigge*. Thus it was not just seeing Ray in my office but what I saw the next day that was such a revelation to me. And one scene, in particular, was so singular that it remains in my memory today as vivid as it was the day I saw it.

My eye was caught by a grey-haired woman in her sixties, who was apparently the center of a most amazing disturbance, though what was happening, what was so disturbing, was not at first clear to me. Was she having a fit? What on earth was convulsing her—and, by a sort of sympathy or contagion, also convulsing everyone whom she gnashingly, ticcily passed?

As I drew closer, I saw what was happening. *She was imitating the passersby*—if "imitation" is not too pallid, too passive, a word. Should we say, rather, that she was caricaturing everyone she passed? Within a second, a split-second, she "had" them all.

I have seen countless mimes and mimics, clowns and antics, but nothing touched the horrible wonder I now beheld: this virtually instantaneous, automatic and convulsive mirroring of every face and figure. But it was not just an imitation, extraordinary as this would have been in itself. The woman not only took on, and took in, the features of countless people, she took them *off*. Every mirroring was also a parody, a mocking, an exaggeration of salient gestures and expressions, but an exaggeration in itself no less convulsive than intentional—a consequence of the violent acceleration and distortion of all her motions. Thus a slow smile, monstrously accelerated, would become a violent, milliseconds-long grimace; an ample gesture, accelerated, would become a farcical convulsive movement.

In the course of a short city block, this frantic old woman frenetically caricatured the features of forty or fifty passersby, in a quick-fire sequence of kaleidoscopic imitations, each lasting a second or two, sometimes less, and the whole dizzying sequence scarcely more than two minutes.

And there were ludicrous imitations of the second and third order; for the people in the street, startled, outraged, bewildered by her imitations, took on these expressions in reaction to her; and those expressions, in turn, were re-reflected, re-directed, re-distorted, by the Touretter, causing a still greater degree of outrage and shock. This grotesque, involuntary resonance or mutuality by which *everyone* was drawn into an absurdly amplifying interaction, was the source of the disturbance I had seen from a distance. This woman who, becoming everybody, lost her own self, became nobody. This woman with a thousand faces, masks, *personae*—how must it be for *her* in this whirlwind of identities? The answer came soon—and not a second too late; for the build-up of pressures, both hers and others', was fast approaching the point of explosion. Suddenly, desperately, the old woman turned aside into an alleyway which led off the main street. And there, with all the appearances of a woman violently sick, she expelled, tremendously accelerated and abbreviated, all the gestures, the postures, the expressions, the demeanors, the entire behavioral repertoires, of the past forty or fifty people she had passed. She delivered one vast, pantomimic egurgitation, in which the engorged identities of the last fifty people who had possessed her were spewed out. And if the taking-in had lasted two minutes, the throwing-out was a single exhalation—fifty people in ten seconds, a fifth of a second or less for the time-foreshortened repertoire of each person.

I was later to spend hundreds of hours talking to, observing, taping, learning from, Tourette patients. Yet nothing, I think, taught me as much, as swiftly, as penetratingly, as overwhelmingly as that phantasmagoric two minutes in a New York street.

It came to me in this moment that such "super-Touretters" must be placed, by an organic quirk, through no fault of their own, in a most extraordinary, indeed unique, existential position, which has some analogies to that of raging "super-Korsakov's," but, of course, has a quite different genesis—and aim. Both can be driven to incoherence, to identity-delirium. The Korsakovian, perhaps mercifully, never knows it, but the Touretter perceives his plight with excruciating and perhaps finally ironic acuity, though he may be unable, or unwilling, to do much about it.

For where the Korsakovian is driven by amnesia, absence, the Touretter is driven by extravagant impulse: impulse of which he is both the creator and the victim, impulse he may repudiate, but cannot disown. Thus he is impelled, as the Korsakovian is not, into an ambiguous relation with his disorder: vanquishing it, being vanquished by it, playing with it—there is every variety of conflict and collusion.

Lacking the normal protective barriers of inhibition, the normal organically determined boundaries of self, the Touretter's ego is subject to a lifelong bombardment. He is beguiled, assailed, by impulses from within and without, impulses which are organic and convulsive but also personal (or rather pseudo-personal) and seductive. How will, how *can*, the ego stand this bombardment? Will identity survive? Can it *develop*, in face of such a shattering, such pressures—or will it be overwhelmed, to produce a "Tourettized soul" (in the poignant words of a patient I was later to see)? There is a physiological, an existential, almost a theological pressure upon the soul of the Touretter—whether it can be held whole and sovereign, or whether it will be taken over, possessed and dispossessed, by every immediacy and impulse.

Hume, as we have noted, wrote:

> I venture to affirm . . . that [we] are nothing but a bundle or collection of different sensations, succeeding one another with inconceivable rapidity, and in a perpetual flux and movement.

Thus, for Hume, personal identity is a fiction—we do not exist, we are but a consecution of sensations, or perceptions.

This is clearly not the case with a normal human being, because he *owns* his own perceptions. They are not a mere flux but *his* own, united by an abiding individuality or self. But what Hume describes may be precisely the case for a being as unstable as a super-Touretter, whose life is, to some extent, a consecution of random or convulsive perceptions and motions, a phantasmagoric fluttering with no center or sense. To this extent he *is* a "Humean" rather than a human being. This is the philosophical, almost theological, fate which lies in wait, if the ratio of impulse to self is too overwhelming. It has affinities to a "Freudian" fate, which is also to be overwhelmed by impulse—but the Freudian fate has sense (albeit tragic), whereas a "Humean" fate is meaningless and absurd.

The super-Touretter, then, is compelled to fight, as no one

else is, simply to survive—to become an individual and survive as one, in face of constant impulse. He may be faced, from earliest childhood, with extraordinary barriers to individuation, to becoming a real person. The miracle is that, in most cases, he succeeds—for the powers of survival, of the will to survive, and to survive as a unique inalienable individual, are absolutely the strongest in our being: stronger than any impulses, stronger than disease. Health, health militant, is usually the victor.

PART THREE

Transports

Introduction

While we have criticized the concept of function, even attempting a rather radical redefinition, we have adhered to it nevertheless, drawing in the broadest terms contrasts based on "deficit" or "excess." But it is clear that wholly other terms also have to be used. As soon as we attend to phenomena as such, to the actual quality of experience or thought or action, we have to use terms more reminiscent of a poem or painting. How, say, is a dream intelligible in terms of function?

We have always two universes of discourse—call them "physical" and "phenomenal," or what you will—one dealing with questions of quantitative and formal structure, the other with those qualities that constitute a "world." All of us have our own distinctive mental worlds, our own inner journeyings and landscapes, and these, for most of us, require no clear neurological "correlate." We can usually tell a man's story, relate passages and scenes from his life, without bringing in any physiological or neurological considerations: such considerations would seem, at the least, supererogatory, if not frankly absurd or insulting. For we consider ourselves, and rightly, "free"—at least, determined by the most complex human and ethical considerations, rather than by the vicissitudes of our neural functions or nervous systems. Usually, but not always: for sometimes a man's life may be cut across, transformed, by

an organic disorder; and if so his story does require a physiological or neurological correlate. This, of course, is so with
all the patients here described.

In the first half of this book we described cases of the obviously pathological—situations in which there is some blatant
neurological excess or deficit. Sooner or later it is obvious
to such patients or their relatives, no less than to their doctors, that there is "something (physically) the matter." Their
inner worlds, their dispositions, may indeed be altered, transformed; but, as becomes clear, this is due to some gross (and
almost quantitative) change in neural function. In this third
section, the presenting feature is reminiscence, altered perception, imagination, "dream." Such matters do not often come
to neurological or medical notice. Such "transports"—often
of poignant intensity, and shot through with personal feeling
and meaning—tend to be seen, like dreams, as psychical: as a
manifestation, perhaps, of unconscious or preconscious activity (or, in the mystically-minded, of something "spiritual"),
not as something "medical," let alone "neurological." They
have an intrinsic dramatic, or narrative, or personal "sense,"
and so are not apt to be seen as "symptoms." It may be in the
nature of transports that they are more likely to be confided to
psychoanalysts or confessors, to be seen as psychoses, or to be
broadcast as religious revelations, rather than brought to physicians. For it never occurs to us at first that a vision might be
"medical"; and if an organic basis is suspected or found, this
may be felt to "devalue" the vision (though, of course, it does
not—values, valuations, have nothing to do with etiology).

All the transports described in this section do have more
or less clear organic determinants (though it was not evident
to begin with but required careful investigation to bring out).
This does not detract in the least from their psychological or
spiritual significance. If God, or the eternal order, was revealed

to Dostoevsky in seizures, why should not other organic conditions serve as "portals" to the beyond or the unknown? In a sense, this section is a study of such portals.

Hughlings Jackson, in 1880, describing such "transports," or "portals," or "dreamy states," in the course of certain epilepsies, used the general word "reminiscence." He wrote:

> I should never diagnose epilepsy from the paroxysmal occurrence of "reminiscence," without other symptoms, although I should suspect epilepsy if that super-positive mental state began to occur very frequently. . . . I have never been consulted for "reminiscence" only.

But *I* have been so consulted: for the forced or paroxysmal reminiscence of tunes, of "visions," of "presences" or scenes—not only in epilepsy but in a variety of other organic conditions. Such transports or reminiscences are not uncommon in migraine (see "The Visions of Hildegard," Chapter 20). This sense of "going back," whether on an epileptic or toxic basis, suffuses "A Passage to India" (Chapter 17). A plainly toxic or chemical basis underlies "Incontinent Nostalgia" (Chapter 16), and the strange hyperosmia of "The Dog Beneath the Skin" (Chapter 18). Either seizure activity or a frontal lobe disinhibition determines the horrifying "reminiscence" of "Murder" (Chapter 19).

The theme of this section is the power of imagery and memory to "transport" a person as a result of abnormal stimulation of the temporal lobes and limbic system of the brain. This may even teach us something of the cerebral basis of certain visions and dreams, and of how the brain (which Sherrington called "an enchanted loom") may weave a magic carpet to transport us.

15

Reminiscence

Mrs. O'C. was somewhat deaf, but otherwise in good health. She lived in an old people's home. One night, in January 1979, she dreamt vividly, nostalgically, of her childhood in Ireland, and especially of the songs they danced to and sang. When she woke up, the music was still going, very loud and clear. "I must still be dreaming," she thought, but this was not so. She got up, roused and puzzled. It was the middle of the night. Someone, she assumed, must have left a radio playing. But why was she the only person to be disturbed by it? She checked every radio she could find—they were all turned off. Then she had another idea: she had heard that dental fillings could sometimes act like a crystal radio, picking up stray broadcasts with unusual intensity. "That's it," she thought. "One of my fillings is playing up. It won't last long. I'll get it fixed in the morning." She complained to the night nurse, who said her fillings looked fine. At this point another notion occurred to Mrs. O'C.: "What sort of radio station," she reasoned to herself, "would play Irish songs, deafeningly, in the middle of the night? Songs, just songs, without introduction or comment? And only songs that I know. What radio station would play *my* songs, and nothing else?" At this point she asked herself, "Is the radio in my head?"

She was now thoroughly rattled—and the music continued deafening. Her last hope was her ENT man, the otologist she was seeing; *he* would reassure her, tell her it was just "noises in the ear," something to do with her deafness, nothing to worry about. But when she saw him in the course of the morning, he said, "No, Mrs. O'C., I don't think it's your ears. A simple ringing or buzzing or rumbling, maybe: but a concert of Irish songs—that's not your ears. Maybe," he continued, "you should see a psychiatrist." Mrs. O'C. arranged to see a psychiatrist the same day. "No, Mrs. O'C.," the psychiatrist said, "it's not your mind. You are not mad—and the mad don't hear music, they only hear 'voices.' You must see a neurologist, my colleague, Dr. Sacks." And so Mrs. O'C. came to me.

Conversation was far from easy, partly because of Mrs. O'C.'s deafness, but more because I was repeatedly drowned out by songs—she could only hear me through the softer ones. She was bright, alert, not delirious or mad, but with a remote, absorbed look, as of someone half in a world of their own. I could find nothing neurologically amiss. Nonetheless, I suspected that the music *was* "neurological."

What could have happened with Mrs. O'C. to bring her to such a pass? She was 88 and in excellent general health with no hint of fever. She was not on any medications which might unbalance her excellent mind. And, manifestly, she had been normal the day before.

"Do you think it's a stroke, Doctor?" she asked, reading my thoughts.

"It could be," I said, "though I've never seen a stroke like this. Something has happened, that's for sure, but I don't think you're in danger. Don't worry, and hold on."

"It's not so easy to hold on," she said, "when you're going through what I'm going through. I know it's quiet here, but I am in an ocean of sound."

I wanted to do an electroencephalogram straightaway,

paying special attention to the temporal lobes, the "musical" lobes of the brain, but circumstances conspired to prevent this for a while. In this time, the music grew less—less loud and, above all, less persistent. She was able to sleep after the first three nights and, increasingly, to make and hear conversation between "songs." By the time I came to do an EEG, she heard only occasional brief snatches of music, a dozen times, more or less, in the course of a day. After we had settled her and applied the electrodes to her head, I asked her to lie still, say nothing and not "sing" to herself, but to raise her right forefinger slightly—which in itself would not disturb the EEG—if she heard any of her songs as we recorded. In the course of a two-hour recording, she raised her finger on three occasions, and each time she did this the EEG pens clattered, and transcribed spikes and sharp waves in the temporal lobes of the brain. This confirmed that she was indeed having temporal lobe seizures, which, as Hughlings Jackson guessed and Wilder Penfield proved, are the invariable basis of "reminiscence" and experiential hallucinations. But why should she suddenly develop this strange symptom? I obtained a brain scan, and this showed that she had indeed had a small thrombosis or infarction in part of her right temporal lobe. The sudden onset of Irish songs in the night, the sudden activation of musical memory-traces in the cortex, were apparently the consequence of a stroke, and as it resolved, so the songs "resolved" too.

By mid-April the songs had entirely gone, and Mrs. O'C. was herself once again. I asked her at this point how she felt about it all, and, in particular, whether she missed the paroxysmal songs she heard. "It's funny you should ask that," she said with a smile. "Mostly, I would say, it is a great relief. But, yes, I *do* miss the old songs a little. Now, with lots of them, I can't even recall them. It was like being given back a forgotten bit of my childhood again. And some of the songs were really lovely."

I had heard similar sentiments from some of my patients on L-Dopa—the term I used was "incontinent nostalgia." And what Mrs. O'C. told me, her obvious nostalgia, put me in mind of a poignant story of H. G. Wells, "The Door in the Wall." I told her the story. "That's it," she said. "That captures the mood, the feeling, entirely. But *my* door is real, as my wall was real. My door leads to the lost and forgotten past."

I did not see a similar case until June last year, when I was asked to see Mrs. O'M., who was now a resident at the same home. Mrs. O'M. was also a woman in her eighties, also somewhat deaf, also bright and alert. She, too, heard music in the head and sometimes a ringing or hissing or rumbling; occasionally she heard "voices talking," usually "far away" and "several at once," so that she could never catch what they were saying. She hadn't mentioned these symptoms to anybody, and had secretly worried, for four years, that she was mad. She was greatly relieved when she heard from the Sister that there had been a similar case in the Home some time before, and very relieved to be able to open up to me.

One day, Mrs. O'M. recounted, while she was grating parsnips in the kitchen, a song started playing. It was "Easter Parade," and was followed, in swift succession, by "Glory, Glory, Hallelujah" and "Good Night, Sweet Jesus." Like Mrs. O'C., she assumed that a radio had been left on, but quickly discovered that all the radios were off. This was in 1979, four years earlier. Mrs. O'C. recovered in a few weeks, but Mrs. O'M.'s music continued, and got worse and worse.

At first she would hear only these three songs—sometimes spontaneously, out of the blue, but for certain if she chanced to think of any of them. She tried, therefore, to avoid thinking of them, but the avoidance of thinking was as provocative as the thinking.

"Do you like these particular songs?" I asked, psychiatrically. "Do they have some special meaning for you?"

"No," she answered promptly. "I never specially liked them, and I don't think they had any special meaning for me."

"And how did you feel when they kept going on?"

"I came to hate them," she replied with great force. "It was like some crazy neighbor continually putting on the same record."

For a year or more, there was nothing but these songs, in maddening succession. After this—and though it was worse in one way, it was also a relief—the inner music became more complex and various. She would hear countless songs, sometimes several simultaneously; sometimes she would hear an orchestra or choir; and, occasionally, voices, or a mere hubbub of noises.

When I came to examine Mrs. O'M. I found nothing abnormal except in her hearing, and here what I found was of singular interest. She had some inner-ear deafness, of a commonplace sort, but over and above this she had a peculiar difficulty in the perception and discrimination of tones of a kind which neurologists call amusia, and which is especially correlated with impaired function in the auditory (or temporal) lobes of the brain. She herself complained that recently the hymns in the chapel seemed more and more alike so that she could scarcely distinguish them by tone or tune but had to rely on the words, or the rhythm.* And although she had been a fine singer in the past, when I tested her she sang flat and out

---

*A similar inability to perceive vocal tone or expression (tonal agnosia) was shown by my patient Emily D. (see "The President's Speech").

of key. She mentioned, too, that her inner music was most vivid when she woke up, becoming less so as other sensory impressions crowded in; and that it was least likely to occur when she was occupied—emotionally, intellectually, but especially visually. In the hour or so she was with me, she heard music only once—a few bars of "Easter Parade," played so loud and so suddenly she could hardly hear me through it.

When we came to do an EEG on Mrs. O'M. it showed strikingly high voltage and excitability in both temporal lobes—those parts of the brain associated with the central representation of sounds and music, and with the evocation of complex experiences and scenes. And whenever she "heard" anything, the high voltage waves became sharp, spike-like, and frankly convulsive. This confirmed my thought that she too had a musical epilepsy, associated with disease of the temporal lobes.

But what *was* going on with Mrs. O'C. and Mrs. O'M.? "Musical epilepsy" sounds like a contradiction in terms: for music, normally, is full of feeling and meaning, and corresponds to something deep in ourselves, "the world behind the music," in Thomas Mann's phrase—whereas epilepsy suggests quite the reverse: a crude, random physiological event, wholly unselective, without feeling or meaning. Thus a "musical epilepsy" or a "personal epilepsy" would seem a contradiction in terms. And yet such epilepsies do occur, though solely in the context of temporal lobe seizures, epilepsies of the reminiscent part of the brain. Hughlings Jackson described these a century ago, and spoke in this context of "dreamy states," "reminiscence," and "physical seizures":

It is not very uncommon for epileptics to have vague and yet exceedingly elaborate mental states at the onset of epileptic seizures. . . . The elaborate mental state, or so-called intellectual aura, is *always the same*, or *essentially the same*, in each case.

Such descriptions remained purely anecdotal until the extraordinary studies of Wilder Penfield, half a century later. Penfield was not only able to locate their origin in the temporal lobes, but was able to *evoke* the "elaborate mental state," or the extremely precise and detailed "experiential hallucinations" of such seizures by gentle electrical stimulation of the seizure-prone points of the cerebral cortex, as this was exposed, at surgery, in fully conscious patients. Such stimulations would instantly call forth intensely vivid hallucinations of tunes, people, scenes, which would be experienced, lived, as compellingly real, in spite of the prosaic atmosphere of the operating room, and could be described to those present in fascinating detail, confirming what Jackson described sixty years earlier, when he spoke of the characteristic "doubling of consciousness":

There is (1) the quasi-parasitical state of consciousness (dreamy state), and (2) there are remains of normal consciousness and thus, there is double consciousness . . . a mental diplopia.

This was precisely expressed to me by my two patients; Mrs. O'M. heard and saw me, albeit with some difficulty, through the deafening dream of "Easter Parade," or the quieter, yet more profound, dream of "Good Night, Sweet Jesus" (which called up for her the presence of a church she used to go to on 31st Street where this was always sung after a novena). And Mrs. O'C. also saw and heard me, through the much profounder anamnestic seizure of her childhood in Ireland: "I know you're there,

Dr. Sacks. I know I'm an old woman with a stroke in an old people's home, but I feel I'm a child in Ireland again—I feel my mother's arms, I see her, I hear her voice singing." Such epileptic hallucinations or dreams, Penfield showed, are never phantasies: they are always memories, and memories of the most precise and vivid kind, accompanied by the emotions which accompanied the original experience. Their extraordinary and consistent detail, which was evoked each time the cortex was stimulated, and exceeded anything which could be recalled by ordinary memory, suggested to Penfield that the brain retained an almost perfect record of every lifetime's experience, that the total stream of consciousness was preserved in the brain, and, as such, could always be evoked or called forth, whether by the ordinary needs and circumstances of life, or by the extraordinary circumstances of an epileptic or electrical stimulation. The variety, the "absurdity," of such convulsive memories and scenes made Penfield think that such reminiscence was essentially meaningless and random:

> At operation it is usually quite clear that the evoked experiential response is a random reproduction of whatever composed the stream of consciousness during some interval of the patient's past life. . . . It may have been [Penfield continues, summarizing the extraordinary miscellany of epileptic dreams and scenes he has evoked] a time of listening to music, a time of looking in at the door of a dance hall, a time of imaging the action of robbers from a comic strip, a time of waking from a vivid dream, a time of laughing conversation with friends, a time of listening to a little son to make sure he was safe, a time of watching illuminated signs, a time of lying in the delivery room at birth, a time of being frightened by a menacing man, a time of watching people enter the room with snow on their clothes. . . . It may have

been a time of standing on the corner of Jacob and Washington, South Bend, Indiana, . . . of watching circus wagons
one night years ago in childhood . . . a time of listening to
(and watching) your mother speed the parting guests . . . or
of hearing your father and mother singing Christmas carols.

I wish I could quote in its entirety this wonderful passage
from Penfield and Perot's massive paper. It gives, as my Irish
ladies do, an amazing feeling of "personal physiology," the
physiology of the self. Penfield is impressed by the frequency
of musical seizures, and gives many fascinating and often
funny examples, a 3 percent incidence in the more than 500
temporal lobe epileptics he has studied:

> We were surprised at the number of times electrical stimu-
> lation has caused the patient to hear *music*. It was produced
> from seventeen different points in 11 cases (see illustration
> on the next page). Sometimes it was an orchestra, at other
> times voices singing, or a piano playing, or a choir. Several
> times it was said to be a radio theme song. . . . The locali-
> sation for production of music is in the superior temporal
> convolution, either the lateral or the superior surface (and,
> as such, close to the point associated with so-called *musi-
> cogenic epilepsy*).

This is borne out, dramatically and often comically, by the
examples Penfield gives. The following list is extracted from
his great final paper:

"White Christmas" (Case 4). Sung by a choir
"Rolling Along Together" (Case 5). Not identified by
    patient, but recognized by operating-room nurse
    when patient hummed it on stimulation

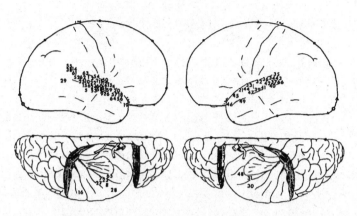

AUDITORY EXPERIENTIAL RESPONSES TO STIMULATION. 1. a voice (14); Case 28. 2. voices (14), 3. a voice (15), 4. a familiar voice (17), 5. a familiar voice (21), 6. a voice (23), 7. a voice (24), 8. a voice (25), 9. a voice (28); Case 29. 10. familiar music (15), 11. a voice (16), 12. a familiar voice (17), 13. a familiar voice (18), 14. familiar music (19), 15. voices (23), 16. voices (27); Case 4. 17. familiar music (14), 18. familiar music (17), 19. familiar music (24), 20. familiar music (25); Case 30. 21. familiar music (23); Case 31. 22. familiar voice (16); Case 32. 23. familiar music (23); Case 5. 24. familiar music (Y), 25. sound of feet walking (1); Case 6. 26. familiar voice (14), 27. voices (22); Case 8. 28. music (15); Case 9. 29. voices (14); Case 36. 30. familiar sound (16); Case 35. 31. a voice (16a); Case 23. 32. a voice (26), 33. voices (25), 34. voices (27), 35. a voice (28), 36. a voice (33); Case 12. 37. music (12); Case 11. 38. a voice (17d); Case 24. 39. familiar voice (14), 40. familiar voices (15), 41. dog barking (17), 42. music (18), 43. a voice (20); Case 13. 44. familiar voice (11), 45. a voice (12), 46. familiar voice (13), 47. familiar voice (14), 48. familiar music (15), 49. a voice (16); Case 14. 50. voices (2), 51. voices (3), 52. voices (5), 53. voices (6), 54. voices (10), 55. voices (11); Case 15. 56. familiar voice (15), 57. familiar voice (16) 58. familiar voice (22); Case 16. 59. music (10); Case 17. 60. familiar voice (30), 61. familiar voice (31), 62. familiar voice (32); Case 3. 63. familiar music (8), 64. familiar music (10), 65. familiar music (D2); Case 10. 66. voices (11); Case 7.

"Hush-a-Bye Baby" (Case 6). Sung by mother, but also thought to be theme-tune for radio-programme

"A song he had heard before, a popular one on the radio" (Case 10)

"Oh Marie, Oh Marie" (Case 30). The theme-song of a radio-programme

"The War March of the Priests" (Case 31). This was on the other side of the "Hallelujah Chorus" on a record belonging to the patient

"Mother and father singing Christmas carols" (Case 32)

"Music from Guys and Dolls" (Case 37)

"A song she had heard frequently on the radio"
   (Case 45)
"I'll Get By" and "You'll Never Know" (Case 46). Songs
 . he had often heard on the radio

In each case—as with Mrs. O'M.—the music was fixed and
stereotyped. The same tune or tunes were heard again and
again, whether in the course of spontaneous seizures, or with
electrical stimulation of the seizure-prone cortex. Thus, these
tunes were not only popular on the radio but equally popular
as hallucinatory seizures; they were, so to speak, the "Top Ten
of the Cortex."

Is there any reason, we must wonder, why particular songs
(or scenes) are "selected" by particular patients for reproduction in their hallucinatory seizures? Penfield considers this
question and feels that there is no reason, and certainly no
significance, in the selection involved:

> It would be very difficult to imagine that some of the trivial
> incidents and songs recalled during stimulation or epilep-
> tic discharge could have any possible emotional signifi-
> cance to the patient, even if one is acutely aware of this
> possibility.

The selection, he concludes, is "quite at random, except
that there is some evidence of cortical conditioning." These
are the words, this is the attitude, so to speak, of physiology.
Perhaps Penfield is right—but could there be more? Is he in
fact "acutely aware," aware enough, at the levels that matter, of the possible emotional significance of songs, of what
Thomas Mann called the "world behind the music"? Would
superficial questioning, such as "Does this song have any spe-

cial meaning for you?" suffice? We know, all too well, from the study of "free associations" that the most seemingly trivial or random thoughts may turn out to have an unexpected depth and resonance, but that this only becomes evident given an analysis in depth. Clearly there is no such deep analysis in Penfield, nor in any other physiological psychology. It is not clear whether any such deep analysis is needed—but given the extraordinary opportunity of such a miscellany of convulsive songs and scenes, one feels, at least, that it should be given a try.

I have gone back to Mrs. O'M. briefly, to elicit her associations, her feelings, to her "songs." This may be unnecessary, but I think it worth trying. One important thing has already emerged. Although, consciously, she cannot attribute to the three songs special feeling or meaning, she now recalls, and this is confirmed by others, that *she was apt to hum them*, unconsciously, long before they became hallucinatory seizures. This suggests that they were *already* unconsciously "selected"—a selection which was then seized on by a supervening organic pathology.

Are they still her favorites? Do they matter to her now? Does she get anything out of her hallucinatory music? The month after I saw Mrs. O'M. there was an article in the *New York Times* entitled "Did Shostakovich Have a Secret?" The "secret" of Shostakovich, it was suggested—by a Chinese neurologist, Dr. Dajue Wang—was the presence of a metallic splinter, a mobile shell fragment, in his brain, in the temporal horn of the left ventricle. Shostakovich was very reluctant, apparently, to have this removed:

Since the fragment had been there, he said, each time he leaned his head to one side he could hear music. His head was filled with melodies—different each time—which he then made use of when composing.

X-rays allegedly showed the fragment moving around when Shostakovich moved his head, pressing against his "musical" temporal lobe when he tilted, producing an infinity of melodies which his genius could use. Dr. R. A. Henson, editor of *Music and the Brain*, expressed deep but not absolute skepticism: "I would hesitate to affirm that it could not happen."

After reading the article, I gave it to Mrs. O'M. to read, and her reactions were strong and clear. "I am no Shostakovich," she said. "I can't use *my* songs. Anyhow, I'm tired of them—they're always the same. Musical hallucinations may have been a gift to Shostakovich, but they are only a nuisance to me. *He* didn't want treatment—but I want it badly."

I put Mrs. O'M. on anticonvulsants, and she forthwith ceased her musical convulsions. I saw her again recently, and asked her if she missed them. "Not on your life," she said. "I'm much better without them." But this, as we have seen, was not the case with Mrs. O'C., whose hallucinosis was of an altogether more complex, more mysterious, and deeper kind and, even if random in its causation, turned out to have great psychological significance and use.

With Mrs. O'C., indeed, the epilepsy was different from the start, both in terms of physiology and of "personal" character and impact. There was, for the first 72 hours, an almost continuous seizure, or seizure "status," associated with an apoplexy of the temporal lobe. This in itself was overwhelming. Secondly, and this too had some physiological basis in the abruptness and extent of the stroke, and its disturbance of deep-lying emotional centers (uncus, amygdala, limbic system, etc., deep within, and deep to the temporal lobe), there

was an overwhelming *emotion* associated with the seizures
and an overwhelming and profoundly nostalgic content—an
overwhelming sense of being a child again, in her long forgotten home, in the arms and presence of her mother.

It may be that such seizures have both a physiological and
a personal origin, coming from particular charged parts of the
brain but, equally, meeting particular psychic circumstances
and needs, as in a case reported by Denis Williams in a 1956
paper:

> A representative, 31 (Case 2770), had major epilepsy
> induced by finding himself alone among strangers. Onset:
> a visual memory of his parents at home, the feeling "How
> marvellous to be back." It is described as a very pleasant
> memory. He gets gooseskin, goes hot and cold, and either
> the attack subsides or proceeds to a convulsion.

Williams relates this astounding story baldly, and makes no
connection between any of its parts. The emotion is dismissed
as purely physiological—inappropriate "ictal pleasure"—and
the possible relation of "being back home" to being lonely is
equally ignored. He may, of course, be right; perhaps it all is
entirely physiological; but I cannot help thinking that if one
has to have seizures, this man, Case 2770, managed to have
the right seizures at the right time.

In Mrs. O'C.'s case the nostalgic need was more chronic
and profound, for her father died before she was born, and her
mother before she was five. Orphaned, alone, she was sent to
America, to live with a rather forbidding maiden aunt. Mrs.
O'C. had no conscious memory of the first five years of her
life—no memory of her mother, of Ireland, of "home." She had
always felt this as a keen and painful sadness—this lack, or

forgetting, of the earliest, most precious years of her life. She had often tried, but never succeeded, to recapture her lost and forgotten childhood memories. Now, with her dream, and the long "dreamy state" which succeeded it, she recaptured a crucial sense of her forgotten, lost childhood. The feeling she had was not just "ictal pleasure" but a trembling, profound and poignant joy. It was, as she said, like the opening of a door—a door which had been stubbornly closed all her life.

In her beautiful book on "involuntary memories," *A Collection of Moments*, Esther Salaman speaks of the necessity to preserve or recapture "the sacred and precious memories of childhood," and how impoverished, *ungrounded*, life is without these. She speaks of the deep joy, the sense of reality, which recapturing such memories may give, and she provides an abundance of marvelous autobiographical quotations, especially from Dostoevsky and Proust. We are all "exiles from our past," she writes, and, as such, we *need* to recapture it. For Mrs. O'C., nearly ninety, approaching the end of a long lonely life, this recapturing of "sacred and precious" childhood memories, this strange and almost miraculous anamnesis, breaking open the closed door, the amnesia of childhood, was provided, paradoxically, by a cerebral mishap.

Unlike Mrs. O'M., who found her seizures exhausting and tiresome, Mrs. O'C. found hers a refreshment to the spirit. They gave her a sense of psychological grounding and reality, the elemental sense which she had lost in her long decades of cut-offness and "exile," that she *had* had a real childhood and home, that she *had* been mothered and loved and cared for. Unlike Mrs. O'M., who *wanted* treatment, Mrs. O'C. declined anticonvulsants. "I *need* these memories," she would say. "I need what's going on. . . . And it'll end by itself soon enough."

Dostoevsky had "psychical seizures" or "elaborate mental states" at the onset of seizures, and once said of these:

> You all, healthy people, can't imagine the happiness which we epileptics feel during the second before our fit. . . . I don't know if this felicity lasts for seconds, hours or months, but believe me, *I would not exchange it for all the joys that life may bring.*

Mrs. O'C. would have understood this. She too knew, in her seizures, an extraordinary felicity. But it seemed to her the acme of sanity and health—the very key, indeed the door, to sanity and health. Thus she felt her illness as health, as *healing.*

As she got better, and recovered from her stroke, Mrs. O'C. had a period of wistfulness and fear. "The door is closing," she said. "I'm losing it all again." And indeed she did lose, by the middle of April, the sudden irruptions of childhood scenes and music and feeling, her sudden epileptic "transports" back to the world of early childhood—which were undoubtedly "reminiscences," and authentic, for, as Penfield has shown beyond doubt, such seizures grasp and reproduce a reality—an experiential reality, and not a phantasy: actual segments of an individual's lifetime and past experience.

But Penfield always speaks of "consciousness" in this regard—of physical seizures as seizing and convulsively replaying part of the stream of consciousness, of conscious reality. What is peculiarly important and moving in the case of Mrs. O'C. is that epileptic "reminiscence" here seized on something unconscious—very early childhood experiences either faded, or repressed from consciousness—and restored them, convulsively, to full memory and consciousness. And it is for this reason, one must suppose, that though, physiologically, the "door" did close, the experience itself was not forgotten but left a profound and enduring impression and was felt as a significant and healing experience. "I'm glad it happened," she said when it was over. "It was the healthiest, happiest experience of my life. There's no longer a great chunk of childhood missing. I can't remember the details now, but I know it's all there. There's a sort of completeness I never had before."

These were not idle words, but brave and true. Mrs. O'C.'s seizures did effect a kind of "conversion," did give a center to a centerless life, did give her back the childhood she had lost—and with this a serenity which she had never had before and which remained for the rest of her life: an ultimate serenity and security of spirit as is only given to those who possess, or recall, the true past.

POSTSCRIPT

"I have never been consulted for 'reminiscence' only," said Hughlings Jackson; in contrast, Freud said, "Neurosis *is* reminiscence." But clearly the word is being used in quite opposite senses—for the aim of psychoanalysis, one might say, is to replace false or fantastic "reminiscences" by a true memory, or anamnesis, of the past (and it is precisely such true memory, trivial or profound, that is evoked in the course of psychical seizures). Freud, we know, greatly admired Jackson—but we do not know if Jackson, who lived to 1911, had ever heard of Freud.

The beauty of a case like Mrs. O'C.'s is that it is at once "Jacksonian" and "Freudian." She suffered from a Jacksonian "reminiscence," but this served to moor and heal her, as a Freudian "anamnesis." Such cases are exciting and precious, for they serve as a bridge between the physical and personal, and they will point, if we let them, to the neurology of the future, a neurology of living experience. This would not, I think, have surprised or outraged Hughlings Jackson. Indeed, it is surely what he himself dreamed of—when he wrote of "dreamy states" and "reminiscence" back in 1880.

Penfield and Perot entitle their paper "The Brain's Record of Visual and Auditory Experience," and we may now meditate on the form or forms such inner "records" may have. What occurs in these wholly personal "experiential" seizures is an entire replay of (a segment of) experience. What, we may ask, *could* be played in such a way as to reconstitute an experience? Is it something akin to a film or record, played on the brain's film projector or phonograph? Or something analogous but logically anterior—such as a script or score? What is the final form, the natural form, of our life's repertoire? That repertoire which provides not only memory and "reminiscence" but our imagination at every level, from the simplest sensory and motor images to the most complex imaginative worlds, landscapes, scenes? A repertoire, a memory, an imagination, of a life which is essentially personal, dramatic and "iconic."

The experiences of reminiscence our patients have raise fundamental questions about the nature of memory (or *mnesis*)—these are also raised, in reverse, in our tales of *a*mnesia or *a*mnesis ("The Lost Mariner" and "A Matter of Identity"). Analogous questions about the nature of knowing (or *gnosis*) are raised by our patients with *a*gnosias—the dramatic visual agnosia of Dr. P. ("The Man Who Mistook His Wife for a Hat"), and the auditory and musical agnosias of Mrs. O'M. or Emily D. (in "The President's Speech"). And similar questions about the nature of action (or *praxis*) are raised by the motor bewilderment, or *a*praxia, of certain retardates, and by patients with frontal lobe apraxias—apraxias which may be so severe that such patients may be unable to walk, may lose their "kinetic melodies," their melodies of walking (this also happens in parkinsonian patients, as was seen in *Awakenings*).

As Mrs. O'C. and Mrs. O'M. suffered from "reminiscence," a convulsive upsurge of melodies and scenes—a sort of *hyper*-mnesis and *hyper*-gnosis—our amnesic-agnosic patients have lost (or are losing) their inner melodies and scenes. Both alike testify to the essentially "melodic" and "scenic" nature of inner life, the Proustian nature of memory and mind.

Stimulate a point in the cortex of such a patient, and there convulsively unrolls a Proustian evocation or reminiscence. What mediates this, we wonder? What sort of cerebral organization could allow this to happen? Our current concepts of cerebral processing and representation are all essentially computational (see, for example, David Marr's brilliant 1982 book, *Vision: A Computational Investigation of Visual Representation in Man*). And, as such, they are couched in terms of "schemata," "programmes," "algorithms," etc.

But could schemata, programs, algorithms alone provide for us the richly visionary, dramatic and musical quality of experience—that vivid personal quality which *makes* it "experience"?

The answer is clearly, even passionately, "No!" Computational representations—even of the exquisite sophistication envisaged by Marr and Bernstein (the two greatest pioneers and thinkers in this realm)—could never, of themselves, constitute "iconic" representations, those representations which are the very thread and stuff of life.

Thus a gulf appears, indeed a chasm, between what we learn from our patients and what physiologists tell us. Is there any way of bridging this chasm? Or, if that is (as it may be) categorically impossible, are there any concepts beyond those of cybernetics by which we may better understand the essentially personal, Proustian nature of reminiscence of the mind, of life? Can we, in short, have a personal or Proustian physiology, over and above the mechanical, Sherringtonian one? Sherrington himself hints at this in *Man on His Nature*, when he imagines the mind as "an enchanted loom," weaving everchanging yet always meaningful patterns—weaving, in effect, patterns of meaning.

Such patterns of meaning would indeed transcend purely formal or computational programs or patterns and allow the essentially *personal* quality which is inherent in reminiscence, inherent in *all* mnesis, gnosis, and praxis. And if we ask what form, what organization, such patterns could have, the answer springs immediately (and, as it were, inevitably) to mind. Personal patterns, patterns for the individual, would have to take the form of scripts or scores—just as abstract patterns, patterns for a computer, must take the form of schemata or programs. Thus, above the level of cerebral programs, we must conceive a level of cerebral scripts and scores.

The score of "Easter Parade," I conjecture, is indelibly inscribed in Mrs. O'M.'s brain—the score, *her* score, of all she heard and felt at the original moment and imprinting of the experience. Similarly, in the "dramaturgic" portions of Mrs. O'C.'s brain, apparently forgotten but nonetheless totally recoverable, must have lain, indelibly inscribed, the script of *her* dramatic childhood scene.

And let us note, from Penfield's cases, that the removal of the minute, convulsing point of cortex, the irritant focus causing reminiscence, can remove *in toto* the iterating scene, and replace an absolutely specific reminiscence or "hyper-mnesia" by an equally specific oblivion or amnesia. There is something extremely important and frightening here: the possibility of a *real* psycho-surgery, a neurosurgery of identity (infinitely finer and more specific than our gross amputations and lobotomies, which may damp or deform the whole character but cannot touch individual experiences).

Experience is not *possible* until it is organized iconically; action is not *possible* unless it is organized iconically. "The brain's record" of everything—everything alive—must be iconic. This is the *final* form of the brain's record, even though the preliminary form may be computational or programmatic. The final form of cerebral representation must be, or allow, "art"—the artful scenery and melody of experience and action.

By the same token, if the brain's representations are damaged or destroyed, as in the amnesias, agnosias, apraxias, their reconstitution (if possible) demands a double approach—an attempt to reconstruct damaged programs and systems—as is being developed, extraordinarily, by Soviet neuropsychology; or a direct approach at the level of inner melodies and scenes (as described in *Awakenings*, *A Leg to Stand On* and several cases in this book, especially "Rebecca" and the introduction to Part Four). Either approach may be used—or both may be used in conjunction—if we are to understand or assist brain-damaged patients: a "systematic" therapy, and an "art" therapy, preferably both.

All of this was hinted at a hundred years ago—in Hughlings Jackson's original 1880 account of "reminiscence"; by Korsakov on amnesia in 1887; and by Freud and Anton on agnosias in the 1890s. Their remarkable insights have been half-forgotten, eclipsed by the rise of a systematic physiology. Now is the time to recall them, re-use them, so that there may arise, in our own time, a new and beautiful "existential" science and therapy, which can join with the systematic, to give us a comprehensive understanding and power.

Since the original publication of this book I have been consulted for innumerable cases of musical "reminiscence"—it is evidently not uncommon, especially in the elderly, though fear may inhibit the seeking of advice. Occasionally (as with Mrs. O'C. and O'M.) a serious or significant pathology is found. Occasionally—as in a recent case report in the *New England Journal of Medicine* (September 5, 1985)—there is a toxic basis, such as the overuse of aspirin. Patients with severe nerve deafness may have musical "phantoms." But in most cases no pathology can be found, and the condition, though a nuisance, is essentially benign. (Why the musical parts of the brain, above all, should be so prone to such "releases" in old age remains far from clear.)

16

Incontinent Nostalgia

If I encountered "reminiscence" occasionally in the context of epilepsy or migraine, I encountered it commonly in my postencephalitic patients excited by L-Dopa—so much so that I found myself calling L-Dopa "a sort of strange and personal time machine." It was so dramatic in one patient that I made her the subject of a letter to the editor, published in the *Lancet* in June 1970. Here I found myself thinking of "reminiscence" in its strict, Jacksonian sense, as a convulsive upsurge of memories from the remote past. Later, when I came to write the history of this patient (Rose R.) in *Awakenings*, I thought less in terms of "reminiscence" and more in terms of "stoppage" ("Has she never moved on from 1926?" I wrote)—and these are the terms in which Harold Pinter portrays "Deborah" in *A Kind of Alaska*. In my letter to the *Lancet*, I wrote:

One of the most astonishing effects of L-Dopa, when given to certain postencephalitic patients, is the reactivation of symptoms and behavior-patterns present at a much earlier stage of the disease, but subsequently "lost." We have already commented, in this connection, on the exacerbation or recurrence of respiratory crises, oculogyric crises, iterative hyperkineses, and tics. We have also observed the reactivation of many other "dormant," primitive symptoms, such as myoclonus, bulimia, polydipsia, satyriasis, central pain, forced affects, etc. At still higher levels of function, we have seen the return and reactivation of elaborate, affectively charged moral postures, thought systems, dreams, and memories—all "forgotten," repressed, or otherwise inactivated in the limbo of profoundly akinetic, and sometimes apathetic, postencephalitic illness.

A striking example of forced reminiscence induced by L-Dopa was seen in the case of a 63-year-old woman who had had progressive postencephalitic parkinsonism since the age of 18 and had been institutionalised, in a state of almost continuous oculogyric "trance," for 24 years. L-Dopa produced, at first, a dramatic release from her parkinsonism and oculogyric entrancement, allowing almost normal speech and movement. Soon there followed (as in several of our patients) a psychomotor excitement with increased libido. This period was marked by nostalgia, joyful identification with a youthful self, and uncontrollable upsurge of remote sexual memories and allusions. The patient requested a tape recorder, and in the course of a few days recorded innumerable salacious songs, "dirty" jokes and limericks, all derived from party gossip, "smutty" comics, nightclubs, and music halls of the middle and late 1920s. These recitals were enlivened by repeated allusions to then-contemporary events, and the use of obsolete colloquialisms, intonations and social mannerisms irresistibly evocative of that bygone flappers' era. Nobody was more astonished than the patient herself: "It's amazing," she said. "I can't understand it. I haven't heard or thought of those things for more than 40 years. I never knew I still knew them. But now they keep running through

my mind." Increasing excitement necessitated a reduction of the dosage of L-Dopa, and with this the patient,
although remaining quite articulate, instantly "forgot"
all these early memories and was never again able to
recall a single line of the songs she had recorded.

Forced reminiscence—usually associated with a
sense of *déjà vu*, and (in Jackson's term) "a doubling of
consciousness"—occurs rather commonly in attacks of
migraine and epilepsy, in hypnotic and psychotic states,
and, less dramatically, in everybody, in response to the
powerful mnemonic stimulus of certain words, sounds,
scenes, and especially smells. Sudden memory-upsurge
has been described as occurring in oculogyric crises, as
in a case described by Zutt in which "thousands of memories suddenly crowded into the patient's mind." Penfield
and Perot have been able to evoke stereotyped recalls
by stimulating epileptogenic points in the cortex, and
surmise that naturally occurring or artificially induced
seizures, occurring in such patients, activate "fossilized
memory sequences" in the brain.

We surmise that our patient (like everybody) is stacked
with an almost infinite number of "dormant" memory
traces, some of which can be reactivated under special conditions, especially conditions of overwhelming
excitement. Such traces, we conceive—like the subcorti-
cal imprints of remote events far below the horizon of
mental life—are indelibly etched in the nervous system,
and may persist indefinitely in a state of abeyance, due
either to lack of excitation or to positive inhibition. The
effects of their excitation or disinhibition may, of course,
be identical and mutually provocative. We doubt, however, whether it is adequate to speak of our patient's

memories as having been simply "repressed" during her illness, and then "depressed" in response to L-Dopa.

The forced reminiscence induced by L-Dopa, cortical probes, migraines, epilepsies, crises, etc., would seem to be, primarily, an excitation; while the incontinently nostalgic reminiscence of old age, and sometimes of drunkenness, seems closer to a disinhibition and uncovering of archaic traces. All of these states can "release" memory, and all of them can lead to a re-experience and re-enactment of the past.

17

A Passage to India

Bhagawhandi P., an Indian girl of 19 with a malignant brain tumor, was admitted to our hospice in 1978. The tumor—an astrocytoma—had first presented when she was seven, but was then of low malignancy and well circumscribed, allowing a complete resection, and complete return of function, and allowing Bhagawhandi to return to normal life.

This reprieve lasted for ten years, during which she lived life to the full, lived it gratefully and consciously to the full, for she knew (she was a bright girl) that she had a "time bomb" in her head.

In her eighteenth year, the tumor recurred, much more invasive and malignant now, and no longer removable. A decompression was performed to allow its expansion—and it was with this, with weakness and numbness of the left side, with occasional seizures and other problems, that Bhagawhandi was admitted.

She was, at first, remarkably cheerful, seeming to accept fully the fate which lay in store, but still eager to be with people and do things, enjoy and experience as long as she could. As the tumor inched forward to her temporal lobe and the decompression started to bulge (we put her on steroids to reduce cerebral edema) her seizures became more frequent—and stranger.

The original seizures were *grand mal* convulsions, and

these she continued to have on occasion. Her new ones had a different character altogether. She would not lose consciousness, but she would look (and feel) "dreamy"; and it was easy to ascertain (and confirm by EEG) that she was now having frequent temporal lobe seizures, which, as Hughlings Jackson taught, are often characterized by "dreamy states" and involuntary "reminiscence."

Soon this vague dreaminess took on a more defined, more concrete, and more visionary character. It now took the form of visions of India—landscapes, villages, homes, gardens—which Bhagawhandi recognized at once as places she had known and loved as a child.

"Do these distress you?" we asked. "We can change the medication."

"No," she said, with a peaceful smile, "I like these dreams—they take me back home."

At times there were people, usually her family or neighbors from her home village; sometimes there was speech, or singing, or dancing; once she was in church, once in a graveyard; but mostly there were the plains, the fields, the rice paddies near her village, and the low, sweet hills which swept up to the horizon.

Were these all temporal lobe seizures? This first seemed the case, but now we were less sure; for temporal lobe seizures (as Hughlings Jackson emphasized, and Wilder Penfield was able by stimulation of the exposed brain to confirm—see "Reminiscence") tend to have a rather fixed format: a single scene or song, unvaryingly reiterated, going with an equally fixed focus in the cortex. Whereas Bhagawhandi's dreams had no such fixity, but presented ever-changing panoramas and dissolving landscapes to her eye. Was she then toxic and hallucinating from the massive doses of steroids she was now receiving?

This seemed possible, but we could not reduce the steroids—she would have gone into coma and died within days.

And a "steroid psychosis," so-called, is often excited and disorganized, whereas Bhagawhandi was always lucid, peaceful and calm. Could they be, in the Freudian sense, phantasies or dreams? Or the sort of dream-madness (oneirophrenia) which may sometimes occur in schizophrenia? Here again we could not be certain; for though there was a phantasmagoria of sorts, yet the phantasms were clearly all memories. They occurred side by side with normal awareness and consciousness (Hughlings Jackson, as we have seen, speaks of a "doubling of consciousness"), and they were not obviously "over-cathected," or charged with passionate drives. They seemed more like certain paintings, or tone poems, sometimes happy, sometimes sad, evocations, revocations, visitations to and from a loved and cherished childhood.

Day by day, week by week, the dreams, the visions, came oftener, grew deeper. They were not occasional now, but occupied most of the day. We would see her rapt, as if in a trance, her eyes sometimes closed, sometimes open but unseeing, and always a faint, mysterious smile on her face. If anyone approached her, or asked her something, as the nurses had to do, she would respond at once, lucidly and courteously, but there was, even among the most down-to-earth staff, a feeling that she was in another world, and that we should not interrupt her. I shared this feeling and, though curious, was reluctant to probe. Once, just once, I said, "Bhagawhandi, what is happening?"

"I am dying," she answered. "I am going home. I am going back where I came from—you might call it my return."

Another week passed, and now Bhagawhandi no longer responded to external stimuli, but seemed wholly enveloped in a world of her own, and, though her eyes were closed, her face still bore its faint, happy smile. "She's on the return journey," the staff said. "She'll soon be there." Three days later she died—or should we say she "arrived," having completed her passage to India?

18

The Dog Beneath the Skin

Stephen D., aged 22, medical student, on highs (cocaine, PCP, chiefly amphetamines).

Vivid dream one night, dreamt he was a dog, in a world unimaginably rich and significant in smells. ("The brilliant smell of water . . . the brave smell of a stone.") Waking, he found himself in just such a world. "As if I had been totally colorblind before, and suddenly found myself in a world full of color." He did, in fact, have an enhancement of color vision ("I could distinguish dozens of browns where I'd just seen brown before. My leatherbound books, which looked similar before, now all had quite distinct and distinguishable hues") and a dramatic enhancement of eidetic visual perception and memory ("I could never draw before, I couldn't 'see' things in my mind, but now it was like having a camera lucida in my mind—I 'saw' everything, as if projected on the paper, and just drew the outlines I 'saw.' Suddenly I could do the most accurate anatomical drawings"). But it was the exaltation of *smell* which really transformed his world: "I had dreamt I was a dog—it was an olfactory dream—and now I awoke to an infinitely redolent world—a world in which all other sensations, enhanced as they were, paled before smell." And with all this there went a sort of trembling, eager emotion, and a strange nostalgia, as of a lost world, half forgotten, half recalled.*

---

*Somewhat similar states—a strange emotionalism; sometimes nostalgia, "reminiscence," and *déjà vu* associated with intense olfactory halluci-

"I went into a scent shop," he continued. "I had never had much of a nose for smells before, but now I distinguished each one instantly—and I found each one unique, evocative, a whole world." He found he could distinguish all his friends— and patients—by smell: "I went into the clinic, I sniffed like a dog, and in that sniff recognized, before seeing them, the twenty patients who were there. Each had his own olfactory physiognomy, a smell-face, far more vivid and evocative, more redolent, than any sight face." He could smell their emotions— fear, contentment, sexuality—like a dog. He could recognize every street, every shop, by smell—he could find his way around New York, infallibly, by smell.

He experienced a certain impulse to sniff and touch everything ("It wasn't really real until I felt it and smelt it") but suppressed this, when with others, lest he seem inappropriate. Sexual smells were exciting and increased—but no more so, he felt, than food smells and other smells. Smell pleasure was intense—smell displeasure, too—but it seemed to him less a world of mere pleasure and displeasure than a whole aesthetic, a whole judgment, a whole new significance, which surrounded him. "It was a world overwhelmingly concrete, of particulars," he said, "a world overwhelming in immediacy, in immediate significance." Somewhat intellectual before, and inclined to reflection and abstraction, he now found thought,

---

nations, are characteristic of "uncinate seizures," a form of temporal lobe epilepsy first described by Hughlings Jackson about a century ago. Usually the experience is rather specific, but sometimes there is a generalized intensification of smell, a hyperosmia. The uncus, phylogenetically part of the ancient "smell-brain" (or rhinencephalon), is functionally associated with the whole limbic system, which is increasingly recognized to be crucial in determining and regulating the entire emotional "tone." Excitation of this, by whatever means, produces heightened emotionalism and an intensification of the senses. The entire subject, with its intriguing ramifications, has been explored in great detail by David Bear.

abstraction, and categorization somewhat difficult and unreal, in view of the compelling immediacy of each experience.

Rather suddenly, after three weeks, this strange transformation ceased—his sense of smell, all his senses, returned to normal; he found himself back, with a sense of mingled loss and relief, in his old world of pallor, sensory faintness, non-concreteness and abstraction. "I'm glad to be back," he said, "but it's a tremendous loss, too. I see now what we give up in being civilized and human. We need the other—the 'primitive'—as well."

Sixteen years have passed—and student days, amphetamine days, are long over. There has never been any recurrence of anything remotely similar. Dr. D. is a highly successful young internist, a friend and colleague of mine in New York. He has no regrets—but he is occasionally nostalgic: "That smell-world, that world of redolence," he exclaims. "So vivid, so real! It was like a visit to another world, a world of pure perception, rich, alive, self-sufficient, and full. If only I could go back sometimes and be a dog again!"

Freud wrote on several occasions of man's sense of smell as being a "casualty," repressed in growing up and civilization with the assumption of an upright posture and the repression of primitive, pre-genital sexuality. Specific (and pathological) enhancements of smell have indeed been reported as occurring in paraphilia, fetishism, and allied perversions and regressions.* But the disinhibition here described seems far more general, and though associated with excitement—probably an amphetamine-induced dopaminergic excitation—was neither

---

*This is well described in a 1932 paper by A. A. Brill, and contrasted with the overall brilliance, the redolence, of the smell-world, in macrosomatic animals (such as dogs), "savages" and children.

specifically sexual nor associated with sexual regression. Similar hyperosmia, sometimes paroxysmal, may occur in excited hyper-dopaminergic states, as with some postencephalitics on L-Dopa and some patients with Tourette's syndrome.

What we see, if nothing else, is the universality of inhibition, even at the most elemental perceptual level: the need to inhibit what Head regarded as primordial and full of feeling-tone, and called "protopathic," in order to allow the emergence of the sophisticated, categorizing, affectless "epicritic."

The need for such inhibition cannot be reduced to the Freudian, nor should its reduction be exalted, romanticized, to the Blakean. Perhaps we need it, as Head implies, that we may be men and not dogs.* And yet Stephen D.'s experience reminds us, like G. K. Chesterton's poem, "The Song of Quoodle," that sometimes we need to be dogs and not men:

> *They haven't got no noses*
> *The fallen sons of Eve . . .*
> *Oh, for the brilliant smell of water,*
> *the brave smell of a stone!*

POSTSCRIPT

I have recently encountered a sort of corollary of this case—a gifted man who sustained a head injury, severely damaging his olfactory tracts (these are very vulnerable in their long course across the anterior fossa) and, in consequence, entirely losing his sense of smell.

He has been startled and distressed at the effects of this:

---

*See Jonathan Miller's 1970 critique of Head, entitled "The Dog Beneath the Skin," in the *Listener*.

"Sense of smell?" he says. "I never gave it a thought. You don't normally give it a thought. But when I lost it—it was like being struck blind. Life lost a good deal of its savor—one doesn't realize how much 'savor' *is* smell. You *smell* people, you *smell* books, you *smell* the city, you *smell* the spring—maybe not consciously, but as a rich unconscious background to everything else. My whole world was suddenly radically poorer."

There was an acute sense of loss, and an acute sense of yearning, a veritable osmalgia: a desire to remember the smell-world to which he had paid no conscious attention, but which, he now felt, had formed the very ground base of life. And then, some months later, to his astonishment and joy, his favorite morning coffee, which had become "insipid," started to regain its savor. Tentatively he tried his pipe, not touched for months, and here too caught a hint of the rich aroma he loved.

Greatly excited—the neurologists had held out no hope of recovery—he returned to his doctor. But after testing him minutely, using a double-blind technique, his doctor said, "No, I'm sorry, there's not a trace of recovery. You still have a total anosmia. Curious though that you should now 'smell' your pipe and coffee."

What seems to be happening—and it is important that it was only the olfactory tracts, not the cortex, which were damaged—is the development of a greatly enhanced olfactory imagery, almost, one might say, a controlled hallucinosis, so that in drinking his coffee or lighting his pipe—situations normally and previously fraught with associations of smell—he is now able to evoke or re-evoke these unconsciously, and with such intensity as to think, at first, that they are "real."

This power—part conscious, part unconscious—has intensified and spread. Now, for example, he snuffs and "smells" the spring. At least he calls up a smell-memory, or smell-picture, so intense that he can almost deceive himself, and deceive others, into believing that he truly smells it.

We know that such a compensation often occurs with the blind and the deaf. We think of the deaf Beethoven and the blinded Prescott. But I have no idea whether it is common with anosmia.

19

Murder

Donald killed his girl while under the influence of PCP. He had, or seemed to have, no memory of the deed—and neither hypnosis nor sodium amytal served to release any. There was, therefore, it was concluded when he stood trial, not a repression of memory but an organic amnesia—the sort of blackout well described with PCP.

The details, manifest on forensic examination, were macabre, and could not be revealed in open court. They were discussed *in camera*—concealed both from the public and from Donald himself. Comparison was made with the acts of violence occasionally committed during temporal lobe or psychomotor seizures. There is no memory of such acts, and perhaps no intention of violence—those who commit them are considered neither responsible nor culpable but are nonetheless committed for their own and others' safety. This was what happened with the unfortunate Donald.

He spent four years in a psychiatric hospital for the criminally insane—despite doubts as to whether he *was* either criminal or insane. He seemed to accept his incarceration with a certain relief—the sense of punishment was perhaps welcome, and there was, he doubtless felt, security in isolation. "I am not fit for society," he would say, mournfully, when questioned.

Security from sudden, dangerous uncontrol—security, and a sort of serenity too. He had always been interested in plants, and this interest, so constructive and so remote from the danger zone of human relation and action, was strongly encouraged in the prison hospital where he now lived. He took over its ragged, untended grounds and created flower gardens, kitchen gardens, gardens of all sorts. He seemed to have achieved a sort of austere equilibrium, in which human relations, human passions, previously so tempestuous, were replaced by a strange calm. Some considered him schizoid, some sane; everyone felt he had achieved a sort of stability. In his fifth year he started to go out on parole, being allowed to leave the hospital on weekend passes. He had been an avid cyclist, and now he again bought a bike. And it was this which precipitated the second act of his strange history.

He was pedaling, fast, as he liked to, down a steep hill when an oncoming car, badly driven, suddenly loomed on a blind turn. Swerving to avoid a head-on collision, he lost control, and was flung violently, headfirst, onto the road.

He sustained a severe head injury: massive bilateral subdural hematomas, which were at once surgically evacuated and drained; and severe contusion of both frontal lobes. He lay in a coma, hemiplegic, for almost two weeks, and then, unexpectedly, he started to recover. And now, at this point, the "nightmares" began.

The returning, the re-dawning, of consciousness was not sweet—it was beset by a hideous agitation and turmoil, in which the half-conscious Donald seemed to be violently struggling, and kept crying, "Oh God!" and "No!" As consciousness grew clearer, so memory, full memory, a now terrible memory, came with it. There were severe neurological problems—left-sided weakness and numbness, seizures, and severe frontal lobe deficits—and with these, with the last of these, something totally new. *The murder, the deed, lost to memory before, now stood before him in vivid, almost hal-*

lucinatory detail. Uncontrollable reminiscence welled up and overwhelmed him—he kept "seeing" the murder, enacting it, again and again. Was this nightmare, was this madness, or was there now "hypermnesis"—a breakthrough of genuine, veridical, terrifyingly heightened memories?

He was questioned in great detail, with the greatest care to avoid any hints or suggestions—and it was very soon clear that what he now showed was a genuine, if uncontrollable, "reminiscence." *He now knew the minutest details of the murder: all the details revealed by forensic examination, but never revealed in open court—or to him.*

All that had been, or seemed, previously lost or forgotten— even in the face of hypnosis or amytal injection—was now recovered and recoverable. More, it was uncontrollable; and still more, completely unbearable. He twice attempted suicide on the neurosurgical unit and had to be heavily tranquilized and forcibly restrained.

What had happened to Donald—what *was* happening with him? That this was a sudden irruption of psychotic phantasy was ruled out by the veridical quality of the reminiscence shown—and even if it were entirely psychotic phantasy, why should it occur now, quite suddenly, unprecedentedly, with his head injury? There was a psychotic, or near psychotic, charge to the memories—they were, in psychiatric parlance, intensely or over-"cathected"—so much so as to drive Donald to incessant thoughts of suicide. But what would be a normal cathexis for such a memory—the sudden emergence, from total amnesia, not of some obscure Oedipal struggle or guilt, but of an actual murder?

Was it possible that, with the loss of frontal lobe integrity, an essential prerequisite for repression had been lost—and that what we now saw was a sudden, explosive, and specific "de-repression"? None of us had ever heard or read of anything quite like this before, although all of us were very familiar with the general disinhibition seen in frontal lobe syndromes—the impulsiveness, the facetiousness, the loquacity, the salacity, the exhibition of an uninhibited, nonchalant, vulgar Id. But this was not the character which Donald now showed. He was not impulsive, unselective, inappropriate, in the least. His character, judgment and general personality were wholly preserved—it was specifically and solely memories and feelings of the murder which now erupted uncontrollably, obsessing and tormenting him.

Was there a specific excitatory or epileptic element involved? Here EEG studies were especially interesting, because it was evident, using special (nasopharyngeal) electrodes, that in addition to the occasional *grand mal* seizures he had, there was an incessant seething, a deep epilepsy, in both temporal lobes, extending down (one might surmise, but it would need implanted electrodes to confirm) into the uncus, the amygdala, the limbic structures—the emotional circuitry which lies deep to the temporal lobes. Penfield and Perot had reported recurrent "reminiscence," or "experiential hallucinations," in some patients with temporal lobe seizures. But most of the experiences or reminiscences which Penfield described were of a somewhat passive sort—hearing music, seeing scenes, being present perhaps, *but present as a spectator, not as an actor.** None of us had heard of such a patient re-experiencing, or

---

*And yet this was not invariably so. In one particularly horrifying, traumatic case, recorded by Penfield, the patient, a girl of twelve, seemed to herself, in every seizure, to be running frantically from a murderous man who was pursuing her with a writhing bag of snakes. This "experiential hallucination" was a precise replay of an actual horrid incident, which had occurred five years before.

rather re-enacting, a *deed*—but this apparently was what was happening with Donald. No clear decision was ever reached.

It remains only to tell the rest of the story. Youth, luck, time, natural healing, superior pre-traumatic function, aided by a Lurianic therapy for frontal lobe "substitution," have allowed Donald, over the years, to make an enormous recovery. His frontal lobe functions now are almost normal. The use of new anticonvulsants, only available in the last few years, have allowed effective control of his temporal lobe seething—and here again, probably, natural recovery has played a part. Finally, with sensitive and supportive regular psychotherapy, the punitive violence of Donald's self-accusing superego has been mitigated, and the gentler scales of the ego now hold court. But the final, the most important, thing is this: that Donald has now returned to gardening. "I feel at peace gardening," he says to me. "No conflicts arise. Plants don't have egos. They can't hurt your feelings." The final therapy, as Freud said, is work and love.

Donald has not forgotten or re-repressed anything of the murder—if, indeed, repression was operative in the first place—but he is no longer obsessed by it; a physiological and moral balance has been struck.

But what of the status of the first lost, then recovered, memory? Why the amnesia—and the explosive return? Why the total blackout and then the lurid flashbacks? What actually happened in this strange, half-neurological drama? All these questions remain a mystery to this day.

20

The Visions of Hildegard

The religious literature of all ages is replete with descriptions of "visions," in which sublime and ineffable feelings have been accompanied by the experience of radiant luminosity (William James speaks of "photism" in this context). It is

"Vision of the Heavenly City." From a manuscript of Hildegard's *Scivias*, written at Bingen about 1180. This figure is a reconstruction from several visions of migrainous origin.

Figure A

Figure B

Figure C

Figure D

Varieties of migraine hallucination represented in the visions of
Hildegard.

In Figure A, the background is formed of shimmering stars set upon
wavering concentric lines. In Figure B, a shower of brilliant stars
(phosphenes) is extinguished after its passage—the succession of
positive and negative scotomas. In Figures C and D, Hildegard depicts
typically migrainous fortification figures radiating from a central point,
which, in the original, is brilliantly luminous and colored.

impossible to ascertain, in the vast majority of cases, whether the experience represents a hysterical or psychotic ecstasy, the effects of intoxication, or an epileptic or migrainous manifestation. A unique exception is provided in the case of Hildegard of Bingen (1098–1180), a nun and mystic of exceptional intellectual and literary powers, who experienced countless visions from earliest childhood to the close of her life, and has left exquisite accounts and figures of these in the two manuscript codices which have come down to us—*Scivias* and *Liber divinorum operum* ("Book of Divine Works").

A careful consideration of these accounts and figures leaves no room for doubt concerning their nature: they were indisputably migrainous, and they illustrate, indeed, many of the varieties of visual aura earlier discussed. Charles Singer, in the course of an extensive essay on Hildegard's visions, selects the following phenomena as most characteristic of them:

> In all a prominent feature is a point or a group of points of light, which shimmer and move, usually in a wave-like manner, and are most often interpreted as stars or flaming eyes [Figure B]. In quite a number of cases one light, larger than the rest, exhibits a series of concentric circular figures of wavering form [Figure A]; and often definite fortification-figures are described, radiating in some cases from a coloured area [Figures C and D]. Often the lights gave that impression of *working*, boiling or fermenting, described by so many visionaries.

Hildegard writes:

> The visions which I saw I beheld neither in sleep, nor in dreams, nor in madness, nor with my carnal eyes, nor with the ears of the flesh, nor in hidden places; but wakeful, alert,

and with the eyes of the spirit and the inward ears, I perceive them in open view and according to the will of God.

One such vision, illustrated by a figure of stars falling and being quenched in the ocean (Figure B), signifies for her "The Fall of the Angels":

> I saw a great star most splendid and beautiful, and with it an exceeding multitude of falling stars which with the star followed southwards. . . . And suddenly they were all annihilated, being turned into black coals . . . and cast into the abyss so that I could see them no more.

Such is Hildegard's allegorical interpretation. Our literal interpretation would be that she experienced a shower of phosphenes in transit across the visual field, their passage being succeeded by a negative scotoma. Visions with fortification figures are represented in her *Zelus Dei* (Figure C) and *Sedens Lucidus* (Figure D), the fortifications radiating from a brilliantly luminous and (in the original) shimmering and colored point. These two visions are combined in a composite vision (first picture), and in this she interprets the fortifications as the *aedificium* of the city of God.

Great rapturous intensity invests the experience of these auras, especially on the rare occasions when a second scotoma follows in the wake of the original scintillation. She wrote:

> The light which I see is not located, but yet is more brilliant than the sun, nor can I examine its height, length or breadth, and I name it "the cloud of the living light." And as sun, moon, and stars are reflected in water, so the writings, sayings, virtues and works of men shine in it before me. . . .

Sometimes I behold within this light another light which I name "the Living Light itself." . . . And when I look upon it every sadness and pain vanishes from my memory, so that I am again as a simple maid and not as an old woman.

Invested with this sense of ecstasy, burning with profound theophorous and philosophical significance, Hildegard's visions were instrumental in directing her towards a life of holiness and mysticism. They provide a unique example of the manner in which a physiological event, banal, hateful, or meaningless to the vast majority of people, can become, in a privileged consciousness, the substrate of a supreme ecstatic inspiration. One must go to Dostoyevsky, who experienced on occasion ecstatic epileptic auras to which he attached momentous significance, to find an adequate historical parallel:

There are moments, and it is only a matter of five or six seconds, when you feel the presence of the eternal harmony . . . a terrible thing is the frightful clearness with which it manifests itself and the rapture with which it fills you. If this state were to last more than five seconds, the soul could not endure it and would have to disappear. During these five seconds I live a whole human existence, and for that I would give my whole life and not think that I was paying too dearly.

PART FOUR

The World of the Simple

The World of the Shining

Introduction

When I started working with retardates several years ago, I thought it would be dismal, and wrote this to Luria. To my surprise, he replied in the most positive terms, and said that there were no patients, in general, more "dear" to him, and that he counted his hours and years at the Institute of Defectology among the most moving and interesting of his entire professional life. He expresses a similar sentiment in the preface to the first of his clinical biographies (*Speech and the Development of Mental Processes in the Child*): "If an author has the right to express feelings about his own work, I must note the warm sense with which I always turn to the material published in this small book."

What is this "warm sense" of which Luria speaks? It is clearly the expression of something emotional and personal—which would not be possible if the defectives did not respond, did not themselves possess very real sensibilities, emotional and personal potentials, whatever their (intellectual) defects. But it is more. It is an expression of scientific interest—of something that Luria considered of quite peculiar scientific interest. What could this be? Something other than "defects" and "defectology," surely, which are of rather limited interest in themselves. What is it, then, that *is* especially interesting in the simple?

It has to do with qualities of mind which are preserved,

even enhanced, so that, though "mentally defective" in some ways, they may be mentally interesting, even mentally complete, in others. Qualities of mind other than the conceptual—this is what we may explore with peculiar clarity in the simple mind (as we may also in the minds of children and "savages"—though, as Clifford Geertz repeatedly emphasizes, these categories must never be equated: savages are neither simple nor children; children have no savage culture; and the simple are neither savages nor children). Yet there are important kinships—and all that Piaget has opened out for us in the minds of children, and Lévi-Strauss in the "savage mind," awaits us, in a different form, in the mind and world of the simple.*

What awaits our study is equally pleasing to the heart and mind, and, as such, especially incites the impulse to Luria's "romantic science."

What is this quality of mind, this disposition, which characterizes the simple and gives them their poignant innocence, transparency, completeness and dignity—a quality so distinctive we must speak of the "world" of the simple (as we speak of the "world" of the child or the savage)?

If we are to use a single word here, it would have to be "concreteness"—their world is vivid, intense, detailed, yet simple, precisely because it *is* concrete: neither complicated, diluted, nor unified, by abstraction.

By a sort of inversion or subversion of the natural order of things, concreteness is often seen by neurologists as a wretched thing, beneath consideration, incoherent, regressed.

---

*All of Luria's early work was done in these three allied domains, his fieldwork with children in primitive communities in Central Asia, and his studies in the Institute of Defectology. Together these launched his lifelong exploration of human imagination.

Thus for Kurt Goldstein, the greatest systematizer of his generation, the mind, man's glory, lies wholly in the abstract and categorical, and the effect of brain damage, any and all brain damage, is to cast him out from this high realm into the almost subhuman swamplands of the concrete. If a man loses the "abstract-categorical attitude" (Goldstein), or "propositional thought" (Hughlings Jackson), what remains is subhuman, of no moment or interest.

I call this an inversion because the concrete is elemental—it is what makes reality "real," alive, personal and meaningful. All of this is lost if the concrete is lost, as we saw in the case of the almost-Martian Dr. P., "the man who mistook his wife for a hat," who fell (in an un-Goldsteinian way) from the concrete *to* the abstract.

Much easier to comprehend, and altogether more natural, is the idea of the preservation of the concrete in brain damage—not regression *to* it, but preservation *of* it, so that the essential personality and identity and humanity, the *being* of the hurt creature, is preserved.

This is what we see in Zazetsky, "the man with a shattered world"—he remains a man, quintessentially a man, with all the moral weight and rich imagination of a man, despite the devastation of his abstract and propositional powers. Here Luria, while seeming to be supporting the formulations of Hughlings Jackson and Goldstein, is, at the same time, turning their significance upside down. Zazetsky is no feeble Jacksonian or Goldsteinian relic, but a man in his full manhood, a man with his emotions and imagination wholly preserved, perhaps enhanced. His world is not "shattered," despite the book's title—it lacks unifying abstractions, but is experienced as an extraordinarily rich, deep and concrete reality.

I believe all this to be true of the simple also—the more so as, having been simple from the start, they have never known, been seduced by, the abstract, but have always experienced reality direct and unmediated, with an elemental and, at times, overwhelming intensity.

We find ourselves entering a realm of fascination and paradox, all of which centers on the ambiguity of the "concrete." In particular, as physicians, as therapists, as teachers, as scientists, we are invited, indeed compelled, towards *an exploration of the concrete*. This is Luria's "romantic science." Both of Luria's great clinical biographies, or "novels," may indeed be seen as explorations of the concrete: its preservation, in the service of reality, in the brain-damaged Zazetsky; its exaggeration, at the expense of reality, in the "supermind" of the Mnemonist.

Classical science has no use for the concrete—it is equated with the trivial in neurology and psychiatry. It needs a "romantic" science to pay it its full due—to appreciate its extraordinary powers . . . and dangers; and in the simple we are confronted with the concrete head-on, the concrete pure and simple, in unreserved intensity.

The concrete can open doors, and it can close them too. It can constitute the portal to sensibility, imagination, depth. Or it can confine the possessor (or the possessed) to meaningless particulars. We see both of these potentials, as it were amplified, in the simple.

Enhanced powers of concrete imagery and memory, nature's compensation for defectiveness in the conceptual and abstract, can tend in quite opposite directions: towards an obsessive preoccupation with particulars, the development of an eidetic imagery and memory, and the mentality of the Performer or "whiz kid" (as occurred with the Mnemonist, and in ancient times, with overcultivation of the concrete "art of memory"*: we see tendencies to this in Martin A. ("A Walking Grove," Chapter 22), in José ("The Autist Artist," Chapter 24), and especially "the Twins" (Chapter 23)—exaggerated, especially in the twins, by the demands of public performance, coupled with their own obsessionalism and exhibitionism.

But of much greater interest, much more human, much more moving, much more "real"—yet scarcely even recognized in scientific studies of the simple (though immediately seen by sympathetic parents and teachers)—is the *proper* use and development of the concrete.

The concrete, equally, may become a vehicle of mystery, beauty, and depth, a path into the emotions, the imagination, the spirit—fully as much as any abstract conception (perhaps indeed more, as Gershom Scholem has argued in his contrasts of the conceptual and the symbolic, or Jerome Bruner in his contrast of the paradigmatic and the narrative). The concrete is readily imbued with feeling and meaning—more readily, perhaps, than any abstract conception. It readily moves into the aesthetic, the dramatic, the comic, the symbolic, the whole wide deep world of art and spirit. *Conceptually*, then, mental defectives may be cripples—but in their powers of concrete and symbolic apprehension they may be fully the equal of any "normal" individual. (This is science, this is romance too. . . .) No one has expressed this more beautifully than Kierkegaard, in the words he wrote on his deathbed. *"Thou plain man!"* (he writes, and I paraphrase slightly). "The symbolism of the Scriptures is something infinitely high . . . but it is not 'high' in a sense that has anything to do with *intellectual* elevation, or with the *intellectual* differences between man and man. . . . No, it is for all . . . for all is this infinite height attainable."

*See Frances Yates's extraordinary book so titled (1966).

A man may be very "low" intellectually, unable to put a key to a door, much less understand the Newtonian laws of motion, wholly unable to comprehend the world *as concepts*— and yet fully able, and indeed gifted, in understanding the world as concreteness, *as symbols*. This is the other side, the almost sublime other side, of the singular creatures, the gifted simpletons, Martin, José, and the Twins.

Yet, it may be said, they are extraordinary and atypical. I therefore start this final section with Rebecca, a wholly "unremarkable" young woman, a simpleton, with whom I worked twelve years ago. I remember her warmly.

21

Rebecca

Rebecca was no child when she was referred to our clinic. She was nineteen, but, as her grandmother said, "just like a child in some ways." She could not find her way around the block, she could not confidently open a door with a key (she could never "see" how the key went, and never seemed to learn). She had left/right confusion, she sometimes put on her clothes the wrong way—inside out, back-to-front, without appearing to notice, or, if she noticed, without being able to get them right. She might spend hours jamming a hand or foot into the wrong glove or shoe—she seemed, as her grandmother said, to have "no sense of space." She was clumsy and ill-coordinated in all her movements—a "klutz," one report said, a "motor moron" another (although when she danced, all her clumsiness disappeared).

Rebecca had a partial cleft palate, which caused a whistling in her speech; short, stumpy fingers, with blunt, deformed nails; and a high, degenerative myopia requiring very thick spectacles—all stigmata of the same congenital condition which had caused her cerebral and mental defects. She was painfully shy and withdrawn, feeling that she was, and had always been, a "figure of fun."

But she was capable of warm, deep, even passionate attachments. She had a deep love for her grandmother, who had brought her up since she was three (when she was orphaned by the death of both parents). She was very fond of nature, and, if she was taken to the city parks and botanic gardens, spent many happy hours there. She was very fond too of stories, though she never learned to read (despite assiduous, and even frantic, attempts), and would implore her grandmother or others to read to her. "She has a hunger for stories," her grandmother said; and fortunately her grandmother loved reading stories and had a fine reading voice which kept Rebecca entranced. And not just stories—poetry too. This seemed a deep need or hunger in Rebecca—a necessary form of nourishment, of reality, for her mind. Nature was beautiful, but mute. It was not enough. She needed the world represented to her in verbal images, in language, and seemed to have little difficulty following the metaphors and symbols of even quite deep poems, in striking contrast to her incapacity with simple propositions and instructions. The language of feeling, of the concrete, of image and symbol, formed a world she loved and, to a remarkable extent, could enter. Though conceptually (and "propositionally") inept, she was at home with poetic language, and was herself, in a stumbling, touching way, a sort of "primitive," natural poet. Metaphors, figures of speech, rather striking similitudes, would come naturally to her, though unpredictably, as sudden poetic ejaculations or allusions. Her grandmother was devout, in a quiet way, and this also was true of Rebecca: she loved the lighting of the Sabbath candles, the benisons and orisons which thread the Jewish day; she loved going to the synagogue, where she too was loved (and seen as a child of God, a sort of innocent, a holy fool); and she fully understood the liturgy, the chants, the prayers, rites and symbols of which the Orthodox service consists. All this was possible for her, accessible to her, loved by her, despite gross perceptual and spatio-temporal problems, and gross impairments in every schematic capacity—she could not count change, the simplest calculations defeated her, she could never learn to read or write, and she would average 60 or less in IQ tests (though doing notably better on the verbal than the performance parts of the test).

Thus she was a "moron," a "fool," a "booby," or had so appeared, and so been called, throughout her whole life, but one with an unexpected, strangely moving, poetic power. Superficially she *was* a mass of handicaps and incapacities, with the intense frustrations and anxieties attendant on these; at this level she was, and felt herself to be, a mental cripple— beneath the effortless skills, the happy capacities, of others; but at some deeper level there was no sense of handicap or incapacity, but a feeling of calm and completeness, of being fully alive, of being a soul, deep and high, and equal to all others. Intellectually, then, Rebecca felt a cripple; spiritually she felt herself a full and complete being.

When I first saw her—clumsy, uncouth, all-of-a-fumble—I saw her merely, or wholly, as a casualty, a broken creature, whose neurological impairments I could pick out and dissect with precision: a multitude of apraxias and agnosias, a mass of sensorimotor impairments and breakdowns, limitations of intellectual schemata and concepts similar (by Piaget's criteria) to those of a child of eight. A poor thing, I said to myself, with perhaps a "splinter skill," a freak gift, of speech; a mere mosaic of higher cortical functions, Piagetian schemata— most impaired.

The next time I saw her, it was all very different. I didn't have her in a test situation, "evaluating" her in a clinic. I wandered outside—it was a lovely spring day—with a few minutes in hand before the clinic started, and there I saw Rebecca sitting on a bench, gazing at the April foliage quietly, with obvious delight. Her posture had none of the clumsiness which had so impressed me before. Sitting there, in a light dress, her face calm and slightly smiling, she suddenly brought to mind one of Chekov's young women—Irene, Anya, Sonya, Nina— seen against the backdrop of a Chekovian cherry orchard. She could have been any young woman enjoying a beautiful spring day. This was my human, as opposed to my neurological, vision.

As I approached, she heard my footsteps and turned, gave me a broad smile, and wordlessly gestured. "Look at the world," she seemed to say. "How beautiful it is." And then there came out, in Jacksonian spurts, odd, sudden, poetic ejaculations: "spring," "birth," "growing," "stirring," "coming to life," "seasons," "everything in its time." I found myself thinking of Ecclesiastes: "To everything there is a season, and a time to every purpose under the heaven. A time to be born, and a time to die; a time to plant, and a time . . ." This was what Rebecca, in her disjointed fashion, was ejaculating—a vision of seasons, of times, like that of the Preacher. "She is an idiot Ecclesiastes," I said to myself. And in this phrase, my two visions of her—as idiot and as symbolist—met, collided and fused. She had done appallingly in the testing—which, in a sense, was designed, like all neurological and psychological testing, not merely to uncover, to bring out deficits, but to decompose her into functions and deficits. She had come apart, horribly, in formal testing, but now she was mysteriously "together" and composed.

Why was she so decomposed before, how could she be so recomposed now? I had the strongest feeling of two wholly different modes of thought, or of organization, or of being. The first schematic—pattern seeing, problem solving—this is what had been tested, and where she had been found so defective, so disastrously wanting. But the tests had given no inkling of anything *but* the deficits, of anything, so to speak, *beyond* her deficits.

They had given me no hint of her positive powers, her ability to perceive the real world—the world of nature, and perhaps of the imagination—as a coherent, intelligible, poetic whole: her ability to see this, think this, and (when she could) live this; they had given me no intimation of her inner world, which clearly *was* composed and coherent, and approached as something other than a set of problems or tasks.

But what was the composing principle which could allow her composure (clearly it was something other than schematic)? I found myself thinking of her fondness for tales, for narrative composition and coherence. Is it possible, I wondered, that this being before me—at once a charming girl and a moron, a cognitive mishap—can *use* a narrative or dramatic mode to compose and integrate a coherent world, in place of the schematic mode, which, in her, is so defective that it simply doesn't work? And as I thought, I remembered her dancing, and how this could organize her otherwise ill-knit and clumsy movements.

Our tests, our approaches, I thought, as I watched her on the bench, enjoying not just a simple but a sacred view of nature—our approach, our "evaluations," are ridiculously inadequate. They only show us deficits, they do not show us powers; they only show us puzzles and schemata, when we need to see music, narrative, play, a being conducting itself spontaneously in its own natural way.

Rebecca, I felt, was complete and intact as "narrative" being, in conditions which allowed her to organize herself in a narrative way; and this was something very important to know, for it allowed one to see her, and her potential, in a quite different fashion from that imposed by the schematic mode.

It was perhaps fortunate that I chanced to see Rebecca in her so-different modes—so damaged and incorrigible in the one, so full of promise and potential in the other—and that she was one of the first patients I saw in our clinic. For what I saw in her, what she showed me, I now saw in them all.

As I continued to see her, she seemed to deepen. Or perhaps she revealed, or I came to respect, her depths more and more. They were not wholly happy depths—no depths ever are—but they were predominantly happy for the greater part of the year.

Then, in November, her grandmother died, and the light, the joy, she had expressed in April now turned into the deepest grief and darkness. She was devastated, but conducted herself with great dignity. Dignity, ethical depth, was added at this time, to form a grave and lasting counterpoint to the light, lyrical self I had especially seen before.

I called on her as soon as I heard the news, and she received me, with great dignity but frozen with grief, in her small room in the now empty house. Her speech was again ejaculated, "Jacksonian," in brief utterances of grief and lamentation. "Why did she have to go?" she cried; and added, "I'm crying for me, not for her." Then, after an interval, "Grannie's all right. She's gone to her Long Home." Long Home! Was this her own symbol, or an unconscious memory of, or allusion to, Ecclesiastes? "I'm so cold," she cried, huddling into herself. "It's not outside, it's winter inside. Cold as death," she added. "She was a part of me. Part of me died with her."

She was complete in her mourning—tragic and complete—there was absolutely no sense of her being then a "mental defective." After half an hour, she unfroze, regained some of her warmth and animation, said: "It is winter. I feel dead. But I know the spring will come again."

The work of grief was slow but successful, as Rebecca, even when most stricken, anticipated. It was greatly helped by a sympathetic and supportive great-aunt, a sister of her Grannie, who now moved into the house. It was greatly helped by the synagogue, and the religious community, above all by the rites of sitting shiva and the special status accorded her as the bereaved one, the chief mourner. It was helped too, perhaps, by her speaking freely to me. And it was helped also, interestingly, by *dreams*, which she related with animation, and which clearly marked *stages* in the grief work (see Peters, 1983).

As I remember her, like Chekov's Nina, in the April sun, so I remember her, etched with tragic clearness, in the dark November of that year, standing in a bleak cemetery in Queens, saying the Kaddish over her grandmother's grave. Prayers and Bible stories had always appealed to her, going with the happy, the lyrical, the "blessing" side of her life. Now, in the funeral prayers, in the 103rd Psalm, and above all in the Kaddish, she found the right and only words for her comfort and lamentation.

During the intervening months between my first seeing her in April and her grandmother's death that November, Rebecca—like all our "clients" (an odious word then becoming fashionable, supposedly less degrading than "patients"), was pressed into a variety of workshops and classes, as part of our Developmental and Cognitive Drive (these too were "in" terms at the time).

It didn't work with Rebecca, it didn't work with most of them. It was not, I came to think, the right thing to do, because what we did was to drive them full tilt upon their limitations, as had already been done, futilely and often to the point of cruelty, throughout their lives.

We paid far too much attention to the defects of our patients, as Rebecca was the first to tell me, and far too little to what was intact or preserved. To use another piece of jargon, we were far too concerned with "defectology," and far too little with "narratology," the neglected and needed science of the concrete.

Rebecca made clear, by concrete illustrations, by her own self, the two wholly different, wholly separate, forms of thought and mind, "paradigmatic" and "narrative" (in Bruner's terminology). And though equally natural and native to the expanding human mind, the narrative comes first, has spiritual priority. Very young children love and demand stories, and can understand complex matters presented as stories, when their powers of comprehending general concepts, paradigms, are almost nonexistent. It is this narrative or symbolic power which gives *a sense of the world*—a concrete reality in the imaginative form of symbol and story—when abstract thought can provide nothing at all. A child follows the Bible before he follows Euclid. Not because the Bible is simpler (the reverse might be said), but because it is cast in a symbolic and narrative mode.

And in this way Rebecca, at nineteen, was still, as her grandmother said, "just like a child." Like a child, but not a child, because she was adult. (The term "retarded" suggests a persisting child, the term "mentally defective" a defective adult; both terms, both concepts, combine deep truth and falsity.)

With Rebecca—and with other defectives allowed, or encouraged in, a personal development—the emotional and narrative and symbolic powers can develop strongly and exuberantly, and may produce (as in Rebecca) a sort of natural poet—or (as in José) a sort of natural artist—while the paradigmatic or conceptual powers, manifestly feeble from the start, grind very slowly and painfully along, and are only capable of a very limited and stunted development.

Rebecca realized this fully—as she had shown it to me so clearly, right from the very first day I saw her, when she spoke of her clumsiness, and of how her ill-composed and ill-organized movements became well-organized, composed and fluent, with music; and when she *showed* me how she herself was composed by a natural scene, a scene with an organic, aesthetic and dramatic unity and sense.

Rather suddenly, after her grandmother's death, she became clear and decisive. "I want no more classes, no more workshops," she said. "They do nothing for me. They do nothing to bring me together." And then, with that power for the apt model or metaphor I so admired, and which was so well developed in her despite her low IQ, she looked down at the office carpet and said:

"I'm like a sort of living carpet. I need a pattern, a design, like you have on that carpet. I come apart, I unravel, unless there's a design."

I looked down at the carpet as Rebecca said this, and found myself thinking of Sherrington's famous image, comparing the brain/mind to an "enchanted loom," weaving patterns ever-dissolving, but always with meaning. I thought: can one have a raw carpet without a design? Could one have the design without the carpet (but this seemed like the smile without the Cheshire cat)? A "living" carpet, as Rebecca was, had to have both—and she especially, with her lack of schematic structure (the warp and woof, the *knit*, of the carpet, so to speak), might indeed unravel without a design (the scenic or narrative structure of the carpet).

"I must have meaning," she went on. "The classes, the odd jobs have no meaning. . . . What I really love," she added wistfully, "is the theater."

We removed Rebecca from the workshop she hated and managed to enroll her in a special theater group. She loved this—it composed her; she did amazingly well: she became a complete person, poised, fluent, with style, in each role. And now if one sees Rebecca on stage—for theater and the theater group soon became her life—one would never even guess that she was mentally defective.

POSTSCRIPT

The power of music, narrative, and drama is of the greatest practical and theoretical importance. One may see this even in the case of idiots, with IQs below 20 and the extremest motor incompetence and bewilderment. Their uncouth movements may disappear in a moment with music and dancing—suddenly, with music, they know how to move. We see how the retarded, unable to perform fairly simple tasks involving perhaps four or five movements or procedures in sequence, can do these perfectly if they work to music—the sequence of movements they cannot hold as schemes being perfectly holdable as music, i.e., embedded in music. The same may be seen, very dramatically, in patients with severe frontal lobe damage and apraxia—an inability to *do* things, to retain the simplest motor sequences and programs, even to walk, despite perfectly preserved intelligence in all other ways. This procedural defect, or motor idiocy, as one might call it, which completely defeats any ordinary system of rehabilitative instruction, vanishes at once if music is the instructor. All this, no doubt, is the rationale, or one of the rationales, of work songs.

What we see, fundamentally, is the power of music to organize—and to do this efficaciously (as well as joyfully!), when abstract or schematic forms of organization fail. Indeed, it is especially dramatic, as one would expect, precisely when no other form of organization will work. Thus music, or any other form of narrative, is essential when working with the retarded or apraxic—schooling or therapy for them must be centered on music or something equivalent. And in drama there is still more—there is the power of *role* to give organization, to confer, while it lasts, an entire personality. The capacity to perform, to play, to *be*, seems to be a "given" in human life, in a way which has nothing to do with intellectual differences. One sees this with infants, one sees it with the senile, and one sees it, most poignantly, with the Rebeccas of this world.

22

A Walking Grove

M artin A., aged 61, was admitted to our Home towards the end of 1983, having become parkinsonian and unable to look after himself any longer. He had had a nearly fatal meningitis in infancy, which caused retardation, impulsiveness, seizures, and some spasticity on one side. He had very limited schooling, but a remarkable musical education— his father was a famous singer at the Met.

He lived with his parents until their death, and thereafter eked out a marginal living as a messenger, a porter, and a short-order cook—whatever he could do before he was fired, as he invariably was, because of his slowness, dreaminess, or incompetence. It would have been a dull and disheartening life, had it not been for his remarkable musical gifts and sensibilities, and the joy this brought him—and others.

He had an amazing musical memory—"I know more than 2,000 operas," he told me on one occasion—although he had never learned or been able to read music. Whether this would have been possible or not was not clear—he had always depended on his extraordinary ear, his power to retain an opera or an oratorio after a single hearing. Unfortunately his voice was not up to his ear—being tuneful, but gruff, with some spastic dysphonia. His innate, hereditary musical gift had clearly survived the ravages of meningitis and brain damage—or had it? Would he have been a Caruso if undamaged?

Or was his musical development, to some extent, a "compensation" for brain damage and intellectual limitations? We shall never know. What is certain is that his father transmitted not only his musical genes, but his own great love for music, in the intimacy of a father-son relationship, and perhaps the specially tender relation of a parent to a retarded child. Martin— slow, clumsy—was loved by his father, and passionately loved him in return; and their love was cemented by their shared love for music.

The great sorrow of Martin's life was that he could not follow his father and be a famous opera and oratorio singer like him—but this was not an obsession, and he found, and gave, much pleasure with what he *could* do. He was consulted, even by the famous, for his remarkable memory, which extended beyond the music itself to all the details of performance. He enjoyed a modest fame as a "walking encyclopedia," who knew not only the music of two thousand operas, but all the singers who had taken the roles in countless performances, and all the details of scenery, staging, dress, and decor. (He also prided himself on a street-by-street, house-by-house, knowledge of New York—and knowing the routes of all its buses and trains.) Thus he was an opera buff, and something of an "idiot savant" too. He took a certain childlike pleasure in all this—the pleasure of such eidetics and freaks. But the real joy—and the only thing that made life supportable—was actual participation in musical events, singing in the choirs at local churches (he could not sing solo, to his grief, because of his dysphonia), especially in the grand events at Easter and Christmas, the *John* and *Matthew Passions*, the *Christmas Oratorio*, the *Messiah*, which he had done for fifty years, boy and man, in the great churches and cathedrals of the city. He had also sung at the old Metropolitan Opera and, when it was pulled down, at the new one in Lincoln Center, discreetly concealed amid the vast choruses of Wagner and Verdi.

At such times—in the oratorios and passions most of all, but also in the humbler church choirs and chorales—as he soared up into the music, Martin forgot that he was "retarded," forgot all the sadness and badness of his life, sensed a great spaciousness enfold him, felt himself both a true man and a true child of God.

Martin's world, his inner world: what sort of a world did he have? He had very little knowledge of the world at large, at least very little living knowledge, and no interest at all. If a page of an encyclopedia or newspaper was read to him, or a map of Asia's rivers or New York's subways shown to him, it was recorded, instantly, in his eidetic memory. But he had no relation to these eidetic recordings—they were "a-centric," to use Richard Wollheim's term, without him, without anyone, or anything, as a living center. There seemed little or no emotion in such memories—no more emotion than there is in a street map of New York—nor did they connect, or ramify, or get generalized, in any way. Thus his eidetic memory—the freak part of him—did not in itself form or convey any sense of a "world." It was without unity, without feeling, without relation to himself. It was physiological, one felt, like a memory core or memory bank, but not part of a real and personal living self.

And yet, even here, there was a single and striking exception, at once his most prodigious, most personal, and most pious deed of memory. He knew by heart *Grove's Dictionary of Music and Musicians*, the immense nine-volume edition published in 1954—indeed he was a "walking Grove." His father was aging and somewhat ailing by then, could no longer sing actively but spent most of his time at home, playing his great collection of vocal records on the phonograph, going through and singing all his scores—which he did with his now thirty-year-old son (in the closest and most affectionate communion of their lives), and reading aloud Grove's dictionary—all six thousand pages of it—which, as he read, was indelibly printed upon his son's limitlessly retentive, if illiterate, cortex. Grove, thereafter, was "heard" *in his father's voice*—and could never be recollected by him without emotion.

Such prodigious hypertrophies of eidetic memory, especially if employed or exploited "professionally," sometimes seem to oust the real self, or to compete with it and impede its development. And if there is no depth, no feeling, there is also no pain in such memories—and so they can serve as an "escape" from reality. This clearly occurred to a great extent in Luria's Mnemonist, and is poignantly described in the last chapter of his book. It obviously occurred to some extent in Martin A., José, and the Twins but was *also*, in each case, used for reality, even "super-reality"—an exceptional, intense, and mystical sense of the world.

Eidetics apart, what of his world generally? It was, in many respects, small, petty, nasty, and dark—the world of a retardate who had been teased and left out as a child, and then hired and fired, contemptuously, from menial jobs, as a man: the world of someone who had rarely felt himself, or felt regarded as, a proper child or man.

He was often childish, sometimes spiteful, and prone to sudden tantrums—and the language he then used was that of a child. "I'll throw a mudpie in your face!" I once heard him scream, and, occasionally, he spat or struck out. He sniffed, he was dirty, he blew snot on his sleeve—he had the look (and doubtless the feelings) at such times of a small, snotty child. These childish characteristics, topped off by his irritating, eidetic showing off, endeared him to nobody. He soon became unpopular in the Home and found himself shunned by many of the residents. A crisis was developing, with Martin regressing weekly and daily, and nobody was quite sure, at first, what to do. It was at first put down to "adjustment difficulties," such as all patients may experience on giving up independent living outside and coming into a "Home." But Sister felt there was something more specific at work—"something gnawing him, a sort of hunger, a gnawing hunger we can't assuage. It's destroying him," she continued. "We have to *do* something."

So, in January, for the second time, I went to see Martin—and found a very different man: no longer cocky, showing off, as before, but obviously pining, in spiritual and a sort of physical pain.

"What is it?" I asked. "What is the matter?"

"I've got to sing," he said hoarsely. "I can't live without it. And it's not just music—I can't pray without it." And then, suddenly, with a flash of his old memory: "'Music, to Bach, was the apparatus of worship,' Grove article on Bach, page 304 . . . I've never spent a Sunday," he continued, more gently, reflectively, "without going to church, without singing in the choir. I first went there with my father, when I was old enough to walk, and I continued going after his death in 1955. *I've got to go,*" he said fiercely. "It'll kill me if I don't."

"And go you shall," I said. "We didn't know what you were missing."

The church was not far from the Home, and Martin was welcomed back—not only as a faithful member of the congregation and the choir, but as the brains and adviser of the choir that his father had been before him.

With this, life suddenly and dramatically changed. Martin had resumed his proper place, as he felt it. He could sing, he could worship, in Bach's music, every Sunday, and also enjoy the quiet authority that was accorded him.

"You see," he told me, on my next visit, without cockiness, but as a simple matter of fact, "they know I know all Bach's liturgical and choral music. I know all the church cantatas— all 202 that Grove lists—and which Sundays and Holy Days they should be sung on. We are the only church in the diocese with a real orchestra and choir, the only one where all of Bach's vocal works are regularly sung. We do a cantata every Sunday—and we are going to do the *Matthew Passion* this Easter!"

I thought it curious and moving that Martin, a retardate, should have this great passion for Bach. Bach seemed so intellectual—and Martin was a simpleton. What I did not realize, until I started bringing in cassettes of the cantatas, and once of the *Magnificat*, when I visited, was that for all his intellectual limitations Martin's musical intelligence was fully up to appreciating much of the technical complexity of Bach; but, more than this—that it wasn't a question of intelligence at all. Bach lived for him, and he lived in Bach.

Martin did, indeed, have "freak" musical abilities—but they were only freak-like if removed from their right and natural context.

What was central to Martin, as it had been central for his father, and what had been intimately shared between them, was always the *spirit* of music, especially religious music, and of the voice as the divine instrument made and ordained to sing, to raise itself in jubilation and praise.

Martin became a different man, then, when he returned to song and church—recovered himself, recollected himself, became real again. The pseudo-persons—the stigmatized retardate, the snotty, spitting boy—disappeared; as did the irritating, emotionless, impersonal eidetic. The real person reappeared, a dignified, decent man, respected and valued now by the other residents.

But the marvel, the real marvel, was to see Martin when he was actually singing, or in communion with music—listening with an intentness which verged on rapture—"a man in his wholeness wholly attending." At such times—it was the same with Rebecca when she acted, or José when he drew, or the Twins in their strange numerical communion—Martin was, in a word, transformed. All that was defective or pathological fell away, and one saw only absorption and animation, wholeness and health.

POSTSCRIPT

When I wrote this piece, and the two succeeding ones, I wrote solely out of my own experience, with almost no knowledge of the literature on the subject, indeed with no knowledge that there *was* a large literature (see, for example, the fifty-two references in Lewis Hill, 1974). I only got an inkling of it, often baffling and intriguing, after "The Twins" was first published, when I found myself inundated with letters and offprints.

In particular, my attention was drawn to a beautiful and detailed 1970 case study by David Viscott. There are many similarities between Martin and his patient Harriet G. In both cases there were extraordinary powers—which were sometimes used in an "a-centric" or life-denying way, sometimes in a life-affirming and creative way: thus, after her father had read it to her, Harriet retained the first three pages of the Boston telephone directory ("and for several years could give any number on these pages on request"); but, in a wholly different and strikingly creative mode, she could compose and improvise in the style of any composer.

It is clear that both—like the Twins (see the next chapter)—could be pushed, or drawn, into the sort of mechanical feats considered typical of "idiot savants"—feats at once prodigious and meaningless; but also that both, when not pushed or drawn in this fashion, showed a consistent seeking after beauty and order. Though Martin has an amazing memory for random, meaningless facts, his real pleasure comes from order and coherence, whether it be the musical and spiritual order of a cantata, or the encyclopedic order of Grove. Both Bach and Grove communicate a *world*. Martin, indeed, has no world *but* music—as is the case with Viscott's patient—but this world is a real world, makes him real, can transform him. This is marvelous to see with Martin, and it was evidently no less so with Viscott's patient, Harriet G. He writes:

> This ungainly, awkward, inelegant lady, this overgrown five-year-old, became absolutely transformed when I asked her to perform for a seminar at Boston State Hospital. She sat down demurely, stared quietly at the keyboard until we all grew silent, and brought her hands slowly to the keyboard and let them rest a moment. Then she nodded her head and began to play with all the feeling and movement of a concert performer. From that moment she was another person.

One speaks of "idiot savants" as if they had an odd "knack" or talent of a mechanical sort, with no real intelligence or understanding. This, indeed, was what I first thought with Martin—and continued to think until I brought in the *Magnificat*. Only then did it finally become clear to me that Martin could grasp the full complexity of such a work, and that it was not just a knack, or a remarkable rote memory at work, but a genuine and powerful musical intelligence. I was particularly interested, therefore, after this book was first published, to receive a fascinating article by L. K. Miller entitled "Developmentally Delayed Musical Savant's Sensitivity to Tonal Structure." Meticulous study of this five-year-old prodigy, with severe mental and other handicaps due to maternal rubella, showed not rote memory of a mechanical sort, but "impressive sensitivity to the rules governing composition, particularly the role of different notes in determining (diatonic) key-structure . . . [implying] implicit knowledge of structural rules in a generative sense: that is, rules not limited to the specific examples provided by one's experience." This, I am convinced, is the case with Martin, too—and one must wonder whether it may not be true of *all* "idiot savants": that they may be truly and creatively intelligent, and not just have a mechanical "knack" in the specific realms—musical, numerical, visual, whatever—in which they excel. It is the *intelligence* of a Martin, a José, the Twins, albeit in a special and narrow area, that finally forces itself on one; and it is this *intelligence* that must be recognized and nurtured.

23

The Twins

When I first met the twins, John and Michael, in 1966 in a state hospital, they were already well known. They had been on radio and television and made the subject of detailed scientific and popular reports.* They had even, I suspected, found their way into science fiction, a little "fictionalized," but essentially as portrayed in the accounts that had been published.†

The twins, who were then twenty-six years old, had been in institutions since the age of seven, variously diagnosed as autistic, psychotic, or severely retarded. Most of the accounts concluded that, as idiots savants go, there was "nothing much to them"—except for their remarkable "documentary" memories of the tiniest visual details of their own experience, and their use of an unconscious, calendrical algorithm that enabled them to say at once on what day of the week a date far in the past or future would fall. This is the view taken by Steven Smith, in his comprehensive and imaginative book, *The Great Mental Calculators*. There have been, to my knowledge, no further studies of the twins since the 1960s, the brief interest they aroused being quenched by the apparent "solution" of the problems they presented.

---

*W. A. Horwitz et al. (1965), Hamblin (1966).
†See Robert Silverberg's novel *Thorns* (1967), notably pp. 11–17.

But this, I believe, is a misapprehension, perhaps a natural enough one in view of the stereotyped approach, the fixed format of questions, the concentration on one "task" or another, with which the original investigators approached the twins, and by which they reduced them—their psychology, their methods, their lives—almost to nothing.

The reality is far stranger, far more complex, far less explicable, than any of these studies suggest, but it is not even to be glimpsed by aggressive formal "testing," or the usual *60 Minutes*-like interviewing of the twins.

Not that any of these studies, or TV performances, is "wrong." They are quite reasonable, often informative, as far as they go, but they confine themselves to the obvious and testable "surface," and do not go to the depths—do not even hint, or perhaps guess, that there are depths below.

One indeed gets no hint of any depths unless one ceases to test the twins, to regard them as "subjects." One must lay aside the urge to limit and test, and get to know the twins—observe them, openly, quietly, without presuppositions, but with a full and sympathetic phenomenological openness, as they live and think and interact quietly, pursuing their own lives, spontaneously, in their singular way. Then one finds there is something exceedingly mysterious at work, powers and depths of a perhaps fundamental sort, which I have not been able to "solve" in the eighteen years that I have known them.

They are, indeed, unprepossessing at first encounter—a sort of Tweedledum and Tweedledee, indistinguishable, mirror images, identical in face, in body movements, in personality, in mind, identical too in their stigmata of brain and tissue damage. They are undersized, with disturbing disproportions in head and hands, high-arched palates, high-arched feet, monotonous squeaky voices, a variety of peculiar tics and mannerisms, and a very high, degenerative myopia requiring glasses so thick that their eyes seem distorted, giving them the appearance of absurd little professors, peering and pointing, with a misplaced, obsessed, and absurd concentration. And this impression is fortified as soon as one quizzes them—or allows them, as they are apt to do, like pantomime puppets, to start spontaneously on one of their "routines."

This is the picture that has been presented in published articles, and on stage—they tend to be "featured" in the annual show in the hospital I work in—and in their not infrequent, and rather embarrassing, appearances on TV.

The "facts," under these circumstances, are established to monotony. The twins say, "Give us a date—any time in the last or next forty thousand years." You give them a date, and almost instantly they tell you what day of the week it would be. "Another date!" they cry, and the performance is repeated. They will also tell you the date of Easter during the same period of 80,000 years. One may observe, though this is not usually mentioned in the reports, that their eyes move and fix in a peculiar way as they do this—as if they were unrolling, or scrutinizing, an inner landscape, a mental calendar. They have the look of "seeing," of intense visualization, although it has been concluded that what is involved is pure calculation.

Their memory for digits is remarkable—and possibly unlimited. They will repeat a number of three digits, of thirty digits, of three hundred digits, with equal ease. This too has been attributed to a "method."

But when one comes to test their ability to calculate—the typical forte of arithmetical prodigies and "mental calculators"—they do astonishingly badly, as badly as their IQs of 60 might lead one to think. They cannot do simple addition or subtraction with any accuracy and cannot even comprehend what multiplication or division means. What is this: "calculators" who cannot calculate, and lack even the most rudimentary powers of arithmetic?

And yet they are called "calendar calculators"—and it has been inferred and accepted, on next to no grounds, that what is involved is not memory at all, but the use of an unconscious algorithm for calendar calculations. When one recollects how even Carl Friedrich Gauss, one of the greatest of mathematicians and of calculators, too, had the utmost difficulty in working out an algorithm for the date of Easter, it is scarcely credible that these twins, incapable of even the simplest arithmetical methods, could have inferred, worked out, and be using such an algorithm. A great many calculators, it is true, *do* have a larger repertoire of methods and algorithms they have worked out for themselves, and perhaps this predisposed W. A. Horwitz et al. to conclude this was true of the twins too. Steven Smith, taking these early studies at face value, comments:

> Something mysterious, though commonplace, is operating here—the mysterious human ability to form unconscious algorithms on the basis of examples.

If this were the beginning and end of it, they might indeed be seen as commonplace, and not mysterious at all—for the computing of algorithms, which can be done well by machine, is essentially mechanical, and comes into the spheres of "problems," but not "mysteries."

And yet, even in some of their performances, their "tricks," there is a quality that takes one aback. They can tell one the weather, and the events, of any day in their lives—any day from about their fourth year on. Their way of talking (well conveyed by the novelist Robert Silverberg in his portrayal of the character Melangio in *Thorns*) is at once childlike, detailed, without emotion. Give them a date, and their eyes roll for a moment, and then fixate; and in a flat, monotonous voice they tell you of the weather, the bare political events they would have heard of, and the events of their own lives— this last often including the painful or poignant anguish of childhood, the contempt, the jeers, the mortifications they endured, but all delivered in an even and unvarying tone, without the least hint of any personal inflection or emotion. Here, clearly, one is dealing with memories that seem of a "documentary" kind, in which there is no personal reference, no personal relation, no living center whatever.

It might be said that personal involvement, emotion, has been edited out of these memories, in the sort of defensive way one may observe in obsessive or schizoid types (and the twins must certainly be considered obsessive and schizoid). But it could be said, equally, and indeed more plausibly, that memories of this kind never *had* any personal character, for this indeed is a cardinal characteristic of eidetic memory such as this.

But what needs to be stressed—and this is insufficiently remarked on by their studiers, though perfectly obvious to a naive listener prepared to be amazed—is the magnitude of the twins' memory, its apparently limitless (if childish and commonplace) extent, and with this the way in which memories are retrieved. And if you ask them how they can hold so much in their minds—a 300-figure digit, or the trillion events of four decades—they say, very simply, "We see it." And "seeing"— "visualizing"—of extraordinary intensity, limitless range, and perfect fidelity, seems to be the key to this. It seems a native physiological capacity of their minds, in a way which has some analogies to that by which A. R. Luria's famous patient, described in *The Mind of a Mnemonist*, "saw," though perhaps the twins lack the rich synesthesia and conscious organization of the Mnemonist's memories. But there is no doubt, in my mind at least, that there is available to the twins a prodigious panorama, a sort of landscape or physiognomy, of all they have ever heard, or seen, or thought, or done, and that in the blink of an eye, externally obvious as a brief rolling and fixation of the eyes, they are able (with the "mind's eye") to retrieve and "see" nearly anything that lies in this vast landscape.

Such powers of memory are most uncommon, but they are hardly unique. We know little or nothing about why the twins or anyone else has them. Is there, then, anything in the twins that is of deeper interest, as I have been hinting? I believe there is.

It is recorded of Sir Herbert Oakley, the nineteenth-century Edinburgh professor of music, that once, taken to a farm, he heard a pig squeak and instantly cried "G sharp!" Someone ran to the piano, and G sharp it was. My own first sight of the "natural" powers, and "natural" mode, of the twins came in a similar, spontaneous, and (I could not help feeling) rather comic, manner.

A box of matches on their table fell and discharged its contents on the floor: "111," they both cried simultaneously; and then, in a murmur, John said "37." Michael repeated this, John said it a third time and stopped. I counted the matches—it took me some time—and there were 111.

"How could you count the matches so quickly?" I asked. "We didn't count," they said. "We *saw* the 111."

Similar tales are told of Zacharias Dase, the number prodigy, who would instantly call out "183" or "79" if a pile of peas was poured out, and indicate as best he could—he was also a dullard—that he did not count the peas, but just "saw" their number, as a whole, in a flash.

"And why did you murmur '37,' and repeat it three times?" I asked the twins. They said in unison, "37, 37, 37, 111."

And this, if possible, I found even more puzzling. That they should *see* 111—"111-ness"—in a flash was extraordinary, but perhaps no more extraordinary than Oakley's "G sharp"—a sort of "absolute pitch," so to speak, for numbers. But they had then gone on to "factor" the number 111—without having any method, without even "knowing" (in the ordinary way) what factors meant. Had I not already observed that they were incapable of the simplest calculations, and didn't "understand" (or seem to understand) what multiplication or division *was*? Yet now, spontaneously, they had divided a compound number into three equal parts.

"How did you work that out?" I said, rather hotly. They indicated, as best they could, in poor, insufficient terms—but perhaps there are no words to correspond to such things—that they did not "work it out," but just "saw" it, in a flash. John made a gesture with two outstretched fingers and his thumb, which seemed to suggest that they had spontaneously *trisected* the number, or that it "came apart" of its own accord, into these three equal parts, by a sort of spontaneous, numerical "fission." They seemed surprised at my surprise—as if *I* were somehow blind; and John's gesture conveyed an extraordinary sense of immediate, *felt* reality. Is it possible, I said to myself, that they can somehow "see" the properties, not in a conceptual, abstract way, but as *qualities*, felt, sensuous, in some immediate, concrete way? And not simply isolated qualities—like "111-ness"—but qualities of relationship? Perhaps in somewhat the same way as Sir Herbert Oakley might have said "a third," or "a fifth."

I had already come to feel, through their "seeing" events and dates, that they could hold in their minds, *did* hold, an immense mnemonic tapestry, a vast (or possibly infinite) landscape in which everything could be seen, either isolated or in relation. It was isolation, rather than a sense of relation, that was chiefly exhibited when they unfurled their implacable, haphazard "documentary." But might not such prodigious powers of visualization—powers essentially concrete, and quite distinct from conceptualization—might not such powers give them the potential of seeing relations, formal relations, relations of form, arbitrary or significant? If they could see "111-ness" at a glance (if they could see an entire "constellation" of numbers), might they not also "see," at a glance—see, recognize, relate and compare, in an entirely sensual and nonintellectual way—enormously complex formations and constellations of numbers? A ridiculous, even disabling power. I thought of what Borges writes about his character Ireneo Funes:

> We, at one glance, can perceive three glasses on a table; Funes [can perceive] all the leaves and tendrils and fruit that make up a grape vine. . . . A circle drawn on a blackboard, a right angle, a lozenge—all these are forms we can fully and intuitively grasp; Ireneo could do the same with the stormy mane of a pony, with a herd of cattle on a hill. . . . I don't know how many stars he could see in the sky.

Could the twins, who seemed to have a peculiar passion and grasp of numbers—could these twins, who had seen "111-ness" at a glance, perhaps see in their minds a numerical "vine," with all the number-leaves, number-tendrils, number-fruit, that made it up? A strange, perhaps absurd, almost impossible thought—but what they had already shown me was so strange as to be almost beyond comprehension. And it was, for all I knew, the merest hint of what they might do.

I thought about the matter, but it hardly bore thinking about. And then I forgot it. Forgot it until a second, spontaneous scene, a magical scene, which I blundered into completely by chance.

This second time they were seated in a corner together, with a mysterious, secret smile on their faces, a smile I had never seen before, enjoying the strange pleasure and peace they now seemed to have. I crept up quietly, so as not to disturb them. They seemed to be locked in a singular, purely numerical, converse. John would say a number—a six-figure number. Michael would catch the number, nod, smile and seem to savor it. Then he, in turn, would say another six-figure number, and now it was John who received, and appreciated it richly. They looked, at first, like two connoisseurs wine-tasting, sharing rare tastes, rare appreciations. I sat still, unseen by them, mesmerized, bewildered.

What were they doing? What on earth was going on? I could make nothing of it. It was perhaps a sort of game, but it had a gravity and an intensity, a sort of serene and meditative and almost holy intensity, which I had never seen in any ordinary game before, and which I certainly had never seen before in the usually agitated and distracted twins. I contented myself with noting down the numbers they uttered—the numbers that manifestly gave them such delight, and which they "contemplated," savored, shared, in communion.

Had the numbers any meaning, I wondered on the way home, had they any "real" or universal sense, or (if any at all) a merely whimsical or private sense, like the secret and silly "languages" brothers and sisters sometimes work out for themselves? And, as I drove home, I thought of Luria's twins—Liosha and Yura—brain damaged, speech-damaged identical twins, and how they would play and prattle with each other in a primitive, babble-like language of their own (Luria and Yudovich, 1959). John and Michael were not even using words or half-words—simply throwing numbers at each other. Were these "Borgesian" or "Funesian" numbers, mere numeric vines, or pony manes, or constellations, private number-forms—a sort of number argot—known to the twins alone?

As soon as I got home I pulled out tables of powers, factors, logarithms, and primes—mementos and relics of an odd, isolated period in my own childhood, when I too was something of a number brooder, a number "see-er," and had a peculiar passion for numbers. I already had a hunch—and now I confirmed it. *All the numbers, the six-figure numbers which the twins had exchanged, were primes*—i.e., numbers that could be evenly divided by no other whole number than itself or one. Had they somehow seen or possessed such a book as mine—or were they, in some unimaginable way, themselves "seeing" primes, in somewhat the same way as they had "seen" 111-ness, or triple 37-ness? Certainly they could not be *calculating* them—they could calculate nothing.

I returned to the ward the next day, carrying the precious book of primes with me. I again found them closeted in their numerical communion, but this time, without saying anything, I quietly joined them. They were taken aback at first, but when I made no interruption, they resumed their "game" of six-figure primes. After a few minutes I decided to join in, and ventured a number, an eight-figure prime. They both turned towards me, then suddenly became still, with a look of intense concentration and perhaps wonder on their faces. There was a long pause—the longest I had ever known them to make, it must have lasted a half-minute or more—and then suddenly, simultaneously, they both broke into smiles.

They had, after some unimaginable internal process of testing, suddenly seen my own eight-digit number as a prime—and this was manifestly a great joy, a double joy, to them; first because I had introduced a delightful new plaything, a prime of an order they had never previously encountered; and, secondly, because it was evident that I had seen what they were doing, that I liked it, that I admired it, and that I could join in myself.

They drew apart slightly, making room for me, a new number playmate, a third in their world. Then John, who always took the lead, thought for a very long time—it must have been at least five minutes, though I dared not move, and scarcely breathed—and brought out a nine-figure number; and after a similar time his twin, Michael, responded with a similar one. And then I, in my turn, after a surreptitious look in my book, added my own rather dishonest contribution, a ten-figure prime I found in my book.

There was again, and for even longer, a wondering, still silence; and then John, after a prodigious internal contemplation, brought out a twelve-figure number. I had no way of checking this, and could not respond, because my own book—which, as far as I knew, was unique of its kind—did not go beyond ten-figure primes. But Michael was up to it, though it took him five minutes—and an hour later the twins were swapping twenty-figure primes, at least I assume this was so, for I had no way of checking it. Nor was there any easy way, in 1966, unless one had the use of a sophisticated computer. And even then, it would have been difficult, for whether one uses

Eratosthenes' sieve, or any other algorithm, there *is* no simple method of calculating primes. *There is no simple method, for primes of this order—and yet the twins were doing it.*

Again I thought of Dase, whom I had read of years before, in F. W. H. Myers's enchanting 1903 book *Human Personality*:

> We know that Dase (perhaps the most successful of such prodigies) was singularly devoid of mathematical grasp. . . .
> Yet he in twelve years made tables of factors and prime numbers for the seventh and nearly the whole of the eighth million—a task which few men could have accomplished, without mechanical aid, in an ordinary lifetime.

He may thus be ranked, Myers concludes, as the only man who has ever done valuable service to Mathematics without being able to cross the Asses' Bridge.

What is not made clear by Myers, and perhaps was not clear, is whether Dase had any method for the tables he made up, or whether, as hinted in his simple "number-seeing" experiments, he somehow "saw" these great primes, as apparently the twins did.

As I observed them, quietly—this was easy to do, because I had an office on the ward where the twins were housed—I observed them in countless other sorts of number games or number communion, the nature of which I could not ascertain or even guess at.

But it seems likely, or certain, that they are dealing with "real" properties or qualities—for the arbitrary, such as random numbers, gives them no pleasure, or scarcely any, at all. It is clear that they must have "sense" in their numbers—in the same way, perhaps, as a musician must have harmony. Indeed, I find myself comparing them to musicians—or to

Martin (Chapter 22), also retarded, who found in the serene and magnificent architectonics of Bach a sensible manifestation of the ultimate harmony and order of the world, wholly inaccessible to him conceptually because of his intellectual limitations.

"Whoever is harmonically composed," writes Sir Thomas Browne, "delights in harmony . . . and a profound contemplation of the First Composer. There is something in it of Divinity more than the ear discovers; it is an Hieroglyphical and shadowed Lesson of the whole World . . . a sensible fit of that harmony which intellectually sounds in the ears of God. . . . The soul . . . is harmonical, and hath its nearest sympathy unto Musick."

Richard Wollheim, in *The Thread of Life*, makes an absolute distinction between calculations and what he calls "iconic" mental states, and he anticipates a possible objection to this distinction:

> Someone might dispute the fact that all calculations are non-iconic on the grounds that, when he calculates, sometimes, he does so by visualising the calculation on a page. But this is not a counter-example. For what is represented in such cases is not the calculation itself, but a representation of it; it is *numbers* that are calculated, but what is visualized are *numerals*, which represent numbers.

Leibniz, on the other hand, makes a tantalizing analogy between numbers and music: "The pleasure we obtain from music comes from *counting*, but counting unconsciously. Music is nothing but unconscious arithmetic."

What, so far as we can ascertain, is the situation with the twins, and perhaps others? Ernst Toch, the composer (his grandson Lawrence Weschler tells me) could readily hold in his mind after a single hearing a very long string of numbers; but he did this by "converting" the string of numbers to a tune—a melody he himself shaped "corresponding" to the numbers. Jedediah Buxton, one of the most ponderous but tenacious calculators of all time, and a man who had a veritable, even pathological, passion for calculation and counting (he would become, in his own words, "drunk with reckoning"), would "convert" music and drama to numbers. "During the dance," a contemporary account of him recorded in 1754, "he fixed his attention upon the number of steps; he declared after a fine piece of musick, that the innumerable sounds produced by the music had perplexed him beyond measure, and he attended even to Mr Garrick only to count the words that he uttered, in which he said he perfectly succeeded."

Here is a pretty, if extreme, pair of examples—the musician who turns numbers into music, and the counter who turns music into numbers. One could scarcely have, one feels, more opposite sorts of mind, or, at least, more opposite modes of mind.*

I believe the twins, who have an extraordinary "feeling" for numbers, without being able to calculate at all, are allied not to Buxton but to Toch in this matter. Except—and this we ordinary people find so difficult to imagine—except that they do not "convert" numbers into music, but actually feel them, in themselves, as "forms," as "tones," like the multitudinous forms that compose nature itself. They are not calculators, and their numeracy is "iconic." They summon up, they dwell

---

*Something comparable to Buxton's mode, which perhaps appears the more "unnatural" of the two, was shown by my patient Miriam H. in *Awakenings* when she had "arithmomanic" attacks.

among, strange scenes of numbers; they wander freely in great landscapes of numbers; they create, dramaturgically, a whole world made of numbers. They have, I believe, a most singular imagination—and not the least of its singularities is that it can imagine only numbers. They do not seem to "operate" with numbers, non-iconically, like a calculator; they "see" them, directly, as a vast natural scene.

And if one asks, are there analogies, at least, to such an "iconicity," one would find this, I think, in certain scientific minds. Dmitri Mendeleev, for example, carried around with him, written on cards, the numerical properties of elements, until they became utterly "familiar" to him—so familiar that he no longer thought of them as aggregates of properties, but (so he tells us) "as familiar faces." He now saw the elements, iconically, physiognomically, as "faces"—faces that related, like members of a family, and that made up, *in toto*, periodically arranged, the whole formal face of the universe. Such a scientific mind is essentially "iconic" and "sees" all nature as faces and scenes, perhaps as music as well. This "vision," this inner vision, suffused with the phenomenal, nonetheless has an integral relation with the physical, and returning it, from the psychical to the physical, constitutes the secondary, or external, work of such science. ("The philosopher seeks to hear within himself the echoes of the world symphony," writes Nietzsche, "and to re-project them in the form of concepts.") The twins, though morons, hear the world symphony, I conjecture, but hear it entirely in the form of numbers.

The soul is "harmonical" whatever one's IQ, and for some, like physical scientists and mathematicians, the sense of harmony, perhaps, is chiefly intellectual. And yet I cannot think of anything intellectual that is not, in some way, also sensible—indeed the very word "sense" always has this double connotation. Sensible, and in some sense "personal" as well, for one cannot feel anything, find anything "sensible," unless it is, in some way, related or relatable to oneself. Thus the mighty architectonics of Bach provide, as they did for Martin A., "an Hieroglyphical and shadowed Lesson of the whole World," but they are also, recognizably, uniquely, dearly, Bach; and this too was felt, poignantly, by Martin A., and related by him to the love he bore his father.

The twins, I believe, have not just a strange "faculty" but a sensibility, a harmonic sensibility, perhaps allied to that of music. One might speak of it, very naturally, as a "Pythagorean" sensibility—and what is odd is not its existence but that it is apparently so rare. One's soul is "harmonical" whatever one's IQ, and perhaps the need to find or feel some ultimate harmony or order is a universal of the mind, whatever its powers, and whatever form it takes. Mathematics has always been called the "queen of sciences," and mathematicians have always felt number as the great mystery, and the world as organized, mysteriously, by the power of number. This is beautifully expressed in the prologue to Bertrand Russell's *Autobiography*:

With equal passion I have sought knowledge. I have wished to understand the hearts of men. I have wished to know why the stars shine. And I have tried to apprehend the Pythagorean power by which number holds sway above the flux.

It is strange to compare these moron twins to an intellect, a spirit, like that of Bertrand Russell. And yet it is not, I think, so far-fetched. The twins live exclusively in a thought-world of numbers. They have no interest in the stars shining, or the hearts of men. And yet numbers for them, I believe, are not "just" numbers, but significances, signifiers whose "significand" is the world.

They do not approach numbers lightly, as most calculators do. They are not interested in, have no capacity for, cannot comprehend, calculations. They are, rather, serene contemplators of number—and approach numbers with a sense of reverence and awe. Numbers for them are holy, fraught with significance. This is their way—as music is Martin's way—of apprehending the First Composer.

But numbers are not just awesome for them, they are friends too—perhaps the only friends they have known in their isolated, autistic lives. This is a rather common sentiment among people who have a talent for numbers—and Steven Smith, while seeing "method" as all-important, gives many delightful examples of it: George Parker Bidder, who wrote of his early number-childhood, "I became perfectly familiar with numbers up to 100; they became as it were my friends, and I knew all their relations and acquaintances"; or the contemporary Shyam Marathe, from India, who remarked, "When I say that numbers are my friends, I mean that I have some time in the past dealt with that particular number in a variety of ways, and on many occasions have found new and fascinating qualities hidden in it. . . . So, if in a calculation I come across a known number, I immediately look to him as a friend."

Hermann von Helmholtz, speaking of musical perception, says that though compound tones *can* be analyzed and broken down into their components, they are normally heard as qualities, unique qualities of tone, indivisible wholes. He speaks here of a "synthetic perception" which transcends analysis, and is the unanalyzable essence of all musical sense. He compares such tones to faces, and speculates that we may recognize them in somewhat the same, personal way. In brief, he half suggests that musical tones, and certainly tunes, are in fact "faces" for the ear and are recognized, felt, immediately as "persons" (or "personeities"), a recognition involving warmth, emotion, personal relation.

So it seems to be with those who love numbers. These too become recognizable as such—in a single, intuitive, personal "I know you!"* The mathematician Wim Klein has put this well: "Numbers are friends for me, more or less. It doesn't mean the same for you, does it—3,844? For you it's just a three and an eight and a four and a four. But I say, 'Hi! 62 squared.'"

I believe the twins, seemingly so isolated, live in a world full of friends, that they have millions, billions, of numbers to which they say "Hi!" and which, I am sure, say "Hi!" back. But none of the numbers is arbitrary—like 62 squared—nor (and this is the mystery) is it arrived at by any of the usual methods, or any method so far as I can make out. The twins seem to employ a direct cognition—like angels. They see, directly, a universe and heaven of numbers. And this, however singular, however bizarre—but what right have we to call it "pathological"?—provides a singular self-sufficiency and serenity to their lives, and one which it might be tragic to interfere with, or break.

This serenity was, in fact, interrupted and broken up ten years later, when it was felt that the twins should be separated—"for

*Particularly fascinating and fundamental problems are raised by the perception and recognition of faces—for there is much evidence that we recognize faces (at least familiar faces) directly and not by any process of piecemeal analysis or aggregation. This, as we have seen, is most dramatically shown in prosopagnosia, in which, as a consequence of a lesion in the right occipital cortex, patients become unable to recognize faces as such, and have to employ an elaborate, absurd, and indirect route, involving a bit-by-bit analysis of meaningless and separate features (see Chapter 1).

their own good," to prevent their "unhealthy communication together," and in order that they could "come out and face the world . . . in an appropriate, socially acceptable way" (as the medical and sociological jargon had it). They were separated, then, in 1977, with results that might be considered as either gratifying or dire. Both have been moved now into "halfway houses," and do menial jobs, for pocket money, under close supervision. They are able to take buses, if carefully directed and given a token, and to keep themselves moderately presentable and clean, though their moronic and psychotic character is still recognizable at a glance.

This is the positive side—but there is a negative side too (not mentioned in their charts, because it was never recognized in the first place). Deprived of their numerical "communion" with each other, and of time and opportunity for any "contemplation" or "communion" at all—they are always being hurried and jostled from one job to another—they seem to have lost their strange numerical power, and with this the chief joy and sense of their lives. But this is considered a small price to pay, no doubt, for their having become quasi-independent and "socially acceptable."

One is reminded somewhat of the treatment meted out to Nadia—an autistic child with a phenomenal gift for drawing. Nadia too was subjected to a therapeutic regime "to find ways in which her potentialities in other directions could be maximised." The net effect was that she started talking—and stopped drawing. Nigel Dennis comments: "We are left with a genius who has had her genius removed, leaving nothing behind but a general defectiveness. What are we supposed to think about such a curious cure?"

It should be added—this is a point dwelt on by F. W. H. Myers, whose consideration of number prodigies opens his chapter on "Genius"—that the faculty is "strange" and may disappear spontaneously, though it is, as often, lifelong. In the case of the twins, of course, it was not just a "faculty" but the personal and emotional center of their lives. And now they are separated, now it is gone, there is no longer any sense or center to their lives.*

POSTSCRIPT

When he was shown the manuscript of this chapter, Israel Rosenfield pointed out that there are other arithmetics, higher and simpler than the "conventional" arithmetic of operations, and wondered whether the twins' singular powers (and limitations) might not reflect their use of such a "modular" arithmetic. In a note to me, he has speculated that modular algorithms of the sort described by Ian Stewart in *Concepts of Modern Mathematics* may explain the twins' calendrical abilities. Rosenfield wrote:

> Their ability to determine the days of the week within an 80,000-year period suggests a rather simple algorithm. One divides the total number of days between "now" and "then" by seven. If there is no remainder, then that date falls on the same day as "now"; if the remainder is one, then that date is one day later; and so on. Notice that modular arithmetic is cyclic: it consists of repetitive patterns. Perhaps the twins were visualizing these patterns,

---

*On the other hand, should this discussion be thought too singular or perverse, it is important to note that in the case of the twins studied by Luria, their separation was essential for their own development, "unlocked" them from a meaningless and sterile babble and bind, and permitted them to develop as healthy and creative people.

either in the form of easily constructed charts, or some kind of "landscape" like the spiral of integers shown on page 30 of Stewart's book.

This leaves unanswered why the twins communicate in primes. But calendar arithmetic requires the prime of seven. And if one is thinking of modular arithmetic in general, modular division will produce neat cyclic patterns *only* if one uses prime numbers. Since the prime number seven helps the twins to retrieve dates, and consequently the events of particular days in their lives, other primes, they may have found, produce similar patterns to those that are so important for their acts of recollection. (When they say about the matchsticks "111–37 three times," note they are taking the prime 37, and multiplying by three.) In fact, only the prime patterns could be "visualized." The different patterns produced by the different prime numbers (for example, multiplication tables) may be the pieces of visual information that they are communicating to each other when they repeat a given prime number. In short, modular arithmetic may help them to retrieve their past, and consequently the patterns created in using these calculations (which only occur with primes) may take on a particular significance for the twins.

By the use of such a modular arithmetic, Ian Stewart points out, one may rapidly arrive at a unique solution in situations that defeat any "ordinary" arithmetic—in particular homing in (by the so-called "pigeon-hole principle") on extremely large and (by conventional methods) incomputable primes.

If such methods, such visualizations, are regarded as algorithms, they are algorithms of a very peculiar sort—organized not algebraically but spatially, as trees, spirals, architectures, "thoughtscapes"—configurations in a formal yet quasi-sensory mental space. I have been excited by Israel Rosenfield's comments, and Ian Stewart's expositions of "higher" (and especially modular) arithmetics, for these seem to promise, if not a "solution," at least a powerful illumination of otherwise inexplicable powers like those of the twins.

Such higher or deeper arithmetics were conceived, in principle, by Gauss in his 1801 *Disquisitiones Arithmeticae*, but they have only been turned to practical realities in recent years. One has to wonder whether there may not be a "conventional" arithmetic (that is, an arithmetic of operations)—often irritating to teacher and student, "unnatural" and hard to learn—and also a deep arithmetic of the kind described by Gauss, which may be truly innate to the brain, as innate as Chomsky's "deep" syntax and generative grammars. Such an arithmetic, in minds like the twins', could be dynamic and almost alive—globular clusters and nebulae of numbers whorling and evolving in an ever-expanding mental sky.

After publication of "The Twins," I received a great deal of communication both personal and scientific. Some dealt with the specific themes of "seeing" or apprehending numbers, some with the sense or significance which might attach to this phenomenon, some with the general character of autistic dispositions and sensibilities and how they might be fostered or inhibited, and some with the question of identical twins. Especially interesting were the letters from parents of such children, the rarest and most remarkable from parents who had themselves been forced into reflection and research and who had succeeded in combining the deepest feeling and involvement with a profound objectivity. In this category were

Clara Claiborne Park and David Park, highly gifted parents of a highly gifted but autistic child. The Parks' child, "Ella," was a talented artist and was also highly gifted with numbers, especially in her earlier years.* She was fascinated by the "order" of numbers, especially primes. This peculiar feel for primes is evidently not uncommon. Clara Park wrote to me of another autistic child she knew, who covered sheets of paper with numbers written down "compulsively." "All were primes," she noted, and added, "They are windows into another world." Later she mentioned an experience with a young autistic man who was also fascinated by factors and primes, and how he instantly perceived these as "special." Indeed, she writes, the word "special" must be used to elicit a reaction:

> "Anything special, Joe, about that number (4875)?"
> Joe: "It's just divisible by 13 and 25."
> Of another (7241): "It's divisible by 13 and 557."
> And of 8741: "It's a prime number."

Park comments: "No one in his family reinforces his primes; they are a solitary pleasure."

It is not clear, in these cases, *how* the answers are arrived at almost instantaneously: whether they are "worked out," "known" (remembered), or—somehow—just "seen." What is clear is the peculiar sense of pleasure and significance attaching to primes. Some of this seems to go with a sense of formal beauty and symmetry, but some with a peculiar associational "meaning" or "potency." This was often called "magical"

---

*Clara Park's book *The Siege* refers to her daughter by the pseudonym Ella, but in subsequent books and papers, she uses her real name, Jessy Park. Jessy herself has now published several collections of her own artwork.

in Ella's case: numbers, especially primes, called up special thoughts, images, feelings, relationships—some almost too "special" or "magical" to be mentioned. This is well described in David Park's 1974 paper.

Kurt Gödel, in a wholly general way, has discussed how numbers, especially primes, can serve as "markers"—for ideas, people, places, whatever; and such a Gödelian marking would pave the way for an "arithmetization" or "numeralization" of the world (see E. Nagel and J. R. Newman, 1958). If this does occur, it is possible that the twins, and others like them, do not merely live in a world *of* numbers, but in a world, in *the* world, *as* numbers, their number meditation or play being a sort of existential meditation—and, if one can understand it, or find the key (as David Park sometimes does), a strange and precise communication too.

24

The Autist Artist

D raw this," I said, and gave José my pocket watch.

He was about 21, said to be hopelessly retarded, and had earlier had one of the violent seizures from which he suffers. He was thin, fragile-looking.

His distraction, his restlessness, suddenly ceased. He took the watch carefully, as if it were a talisman or jewel, laid it before him, and stared at it in motionless concentration.

"He's an idiot," the attendant broke in. "Don't even ask him. He don't know what it is—he can't tell time. He can't even talk. They say he's 'autistic,' but he's just an idiot." José turned pale, perhaps more at the attendant's tone than at his words—the attendant had said earlier that José didn't use words.

"Go on," I said. "I know you can do it."

José drew with an absolute stillness, concentrating completely on the little clock before him, everything else shut out. Now, for the first time, he was bold, without hesitation, composed, not distracted. He drew swiftly but minutely, with a clear line, without erasures.

I nearly always ask patients, if it is possible for them, to write and draw, partly as a rough-and-ready index of various competences, but also as an expression of "character" or "style."

José had drawn the watch with remarkable fidelity, putting in every feature (at least every essential feature: he did not

put in "Westclox, shock resistant, made in USA"), and not just "the time" (though this was faithfully registered as 11:31) but every second as well, and the inset seconds dial—and, not least, the knurled winder and trapezoid clip of the watch used to attach it to a chain. The clip was strikingly amplified, though everything else remained in due proportion. And the figures, now that I came to look at them, were of different sizes, different shapes, different styles—some thick, some thin; some aligned, some inset; some plain and some elaborated, even a bit "gothic." And the inset second hand, rather inconspicuous in the original, had been given a striking prominence, like the small inner dials of star clocks or astrolabes.

The general grasp of the thing, its "feel," had been strikingly brought out—all the more strikingly if, as the attendant said, José had no idea of time. And otherwise there was an odd mixture of close, even obsessive, accuracy, with curious (and, I felt, droll) elaborations and variations.

I was puzzled by this, haunted by it as I drove home. An "idiot"? Autism? No. Something else was going on here.

I was not called to see José again. The first call, on a Sunday evening, had been for an emergency. He had been having seizures the entire weekend, and I had prescribed changes in his anticonvulsants, over the phone, in the afternoon. Now that his seizures were "controlled," further neurological advice was not requested. But I was still troubled by the problems presented by the clock and felt an unresolved sense of mystery about it. I needed to see him again. So I arranged a further visit, and to see his entire chart—I had been given only a consultation slip, not very informative, when I saw him before.

José came casually into the clinic—he had no idea (and perhaps did not care) why he'd been called—but his face lit up with a smile when he saw me. The dull, indifferent look, the mask I remembered, was lifted. There was a sudden, shy smile, like a glimpse through a door.

"I have been thinking about you, José," I said. He might not understand my words, but he understood my tone. "I want to see more drawing"—and I gave him my pen.

What should I ask him to draw this time? I had, as always, a copy of *Arizona Highways* with me, a richly illustrated magazine which I especially delight in, and which I carry around for neurological purposes, for testing my patients. The cover depicted an idyllic scene of people canoeing on a lake, against a backdrop of mountains and sunset. José started with the foreground, a mass of near-black silhouetted against the water, outlined this with extreme accuracy, and started to block it in. But this was clearly a job for a paintbrush, not a fine pen. "Skip it," I said, then pointing, "Go on to the canoe." Rapidly, unhesitatingly, José outlined the silhouetted figures and the canoe. He looked at them, then looked away, their forms fixed in his mind—then swiftly blocked them in with the side of the pen.

Here again, and more impressively, because an entire scene was involved, I was amazed at the swiftness and the minute

accuracy of reproduction, the more so since José had gazed at the canoe and then away, having taken it in. This argued strongly against any mere copying—the attendant had said earlier, "He's just a Xerox"—and suggested that he had apprehended it as an image, exhibiting a striking power not just of copying but of perception. For the image had a dramatic quality not present in the original. The tiny figures, enlarged, were more intense, more alive, had a feeling of involvement and purpose not at all clear in the original. All the hallmarks of what Richard Wollheim calls "iconicity"—subjectivity, intentionality, dramatization—were present. Thus, over and above the powers of mere facsimile, striking as these were, he seemed to have clear powers of imagination and creativity. It was not *a* canoe but *his* canoe that emerged in the drawing.

I turned to another page in the magazine, to an article on trout fishing, with a pastel watercolor of a trout stream, a background of rocks and trees, and in the foreground a rainbow trout about to take a fly. "Draw this," I said, pointing to

the fish. He gazed at it intently, seemed to smile to himself, and then turned away—and now, with obvious enjoyment, his smile growing broader and broader, he drew a fish of his own.

I smiled myself, involuntarily, as he drew it, because now, feeling comfortable with me, he was letting himself go, and what was emerging, slyly, was not just a fish, but a fish with a "character" of sorts.

The original had lacked character, had looked lifeless, two-dimensional, even stuffed. José's fish, by contrast, tilted and poised, was richly three-dimensional, far more like a real fish than the original. It was not only verisimilitude and anima-tion that had been added but something else, something richly expressive, though not wholly fishlike: a great, cavernous, whalelike mouth; a slightly crocodilian snout; an eye, one

had to say, which was distinctly human, and with altogether a positively roguish look. It was a very funny fish—no wonder he had smiled—a sort of fish-person, a nursery character, like the frog footman in *Alice*.

Now I had something to go on. The picture of the clock had startled me, stimulated my interest, but did not, in itself, allow any thoughts or conclusions. The canoe had shown that José had an impressive visual memory, and more. The fish showed a lively and distinctive imagination, a sense of humor, and something akin to fairy-tale art. Certainly not great art, it was "primitive," perhaps it was child-art; but, without doubt, it was art of a sort. And imagination, playfulness, and art are precisely what one does not expect in idiots, or idiots savants, or in the autistic either. Such at least is the prevailing opinion.

My friend and colleague Isabelle Rapin had actually seen José years before, when he was presented with "intractable seizures" in the child neurology clinic—and she, with her great experience, did not doubt that he was "autistic." Of autism in general she had written:

A small number of autistic children are exceedingly proficient at decoding written language and become hyperlexic or preoccupied with numbers. . . . Extraordinary proficiencies of some autistic children for putting together puzzles, taking apart mechanical toys, or decoding written texts may reflect the consequences of attention and learning being inordinately focused on non-verbal visual-spatial tasks to the exclusion of, or perhaps because of, the lack of demand for learning verbal skills. (1982, pp. 146–50)

Somewhat similar observations, specifically about drawing, are made by Lorna Selfe in her astonishing book *Nadia*.

All idiot savant or autistic proficiencies and performances, Dr. Selfe gathered from the literature, were apparently based on calculation and memory alone, never on anything imaginative or personal. And if these children could draw—supposedly a very rare occurrence—their drawings too were merely mechanical. "Isolated islands of proficiency" and "splinter skills" are spoken of in the literature. No allowance is made for an individual, let alone a creative, personality.

What then was José, I had to ask myself. What sort of being? What went on inside him? How had he arrived at the state he was in? And what state was it—and might anything be done?

I was both assisted and bewildered by the available information—the mass of "data" that had been gathered since the first onset of his strange illness, his "state." I had a lengthy chart available to me, containing early descriptions of his original illness: a very high fever at the age of eight, associated with the onset of incessant, continuing seizures, and the rapid appearance of a brain-damaged or autistic condition. (There had been doubt from the start about what, exactly, was going on.)

His spinal fluid had been abnormal during the acute stage of the illness. The consensus was that he had probably suffered an encephalitis of sorts. His seizures were of many different types—petit mal, grand mal, "akinetic," and "psychomotor," these last being seizures of an exceptionally complex type.

Psychomotor seizures can also be associated with sudden passion and violence, and the occurrence of peculiar behavior states even between seizures (the so-called psychomotor personality). They are invariably associated with disorder in, or damage to, the temporal lobes, and severe temporal lobe disorder, both left-sided and right-sided, had been demonstrated in José by innumerable EEGs.

The temporal lobes are also associated with the auditory capacities, and, in particular, the perception and production of speech. Dr. Rapin had not only considered José "autistic" but had wondered whether a temporal lobe disorder had caused a "verbal auditory agnosia"—an inability to recognize speech sounds that interfered with his capacity to use or understand the spoken word. For what was striking, however it was to be interpreted (and both psychiatric and neurological interpretations were offered), was the loss or regression of speech, so that José, previously "normal" (or so his parents avowed), became "mute" and ceased talking to others when he became ill.

One capacity was apparently "spared"—perhaps in a compensatory way enhanced: an unusual passion and power to draw, which had been evident since early childhood, and seemed to some extent hereditary or familial, for his father had always been fond of sketching, and his (much) older brother was a successful artist. With the onset of his illness, with his seemingly intractable seizures (he might have twenty or thirty major convulsions a day, and uncounted "little seizures," falls, "blanks," or "dreamy states"); with the loss of speech and his general intellectual and emotional "regression," José found himself in a strange and tragic state. His schooling was discontinued, though a private tutor was provided for a while, and he was returned permanently to his family, as a "full-time" epileptic, autistic, perhaps aphasic, retarded child. He was considered ineducable, untreatable and generally hopeless. At the age of nine, he "dropped out"—out of school, out of society, out of almost all of what for a normal child would be "reality."

For fifteen years he scarcely emerged from the house, ostensibly because of "intractable seizures," his mother maintaining she dared not take him out, otherwise he would have twenty

or thirty seizures in the street every day. All sorts of anticonvulsants were tried, but his epilepsy seemed "untreatable":
this, at least, was the stated opinion in his chart. There were
older brothers and sisters, but José was much the youngest—
the "big baby" of a woman approaching fifty.

We have far too little information about these intervening
years. José, in effect, disappeared from the world, was "lost to
follow-up," not only medically but generally, and might have
been lost forever, confined and convulsing in his cellar room,
had he not "blown up" violently very recently and been taken
to the hospital for the first time. He was not entirely without inner life, in the cellar. He showed a passion for pictorial magazines, especially of natural history, of the *National
Geographic* type, and when he was able, between seizures and
scoldings, would find stumps of pencil and draw what he saw.

These drawings were perhaps his only link with the outside world, and especially the world of animals and plants, of
nature, which he had so loved as a child, especially when he
went out sketching with his father. This, and this only, he was
permitted to retain, his one remaining link with reality.

This, then, was the tale I received, or, rather, put together
from his chart or charts, documents as remarkable for what
they lacked as for what they contained—the documentation,
through default, of a fifteen-year "gap": from a social worker
who had visited the house, taken an interest in him, but could
do nothing; and from his now aged and ailing parents as well.
But none of this would have come to light had there not been
a rage of sudden, unprecedented, and frightening violence—a
fit in which objects were smashed—which brought José to a
state hospital for the first time.

It was far from clear what had caused this rage, whether
it was an eruption of epileptic violence (such as one may see,

on rare occasions, with very severe temporal lobe seizures), or whether it was, in the simplistic terms of his admission note, simply "a psychosis," or whether it represented some final, desperate call for help from a tortured soul who was mute and had no direct way of expressing his predicament, his needs.

What was clear was that coming to the hospital and having his seizures controlled by powerful new drugs for the first time gave him some space and freedom, a release both physiological and psychological, of a sort he had not known since the age of eight.

Hospitals, state hospitals, are often seen as "total institutions" in Erving Goffman's sense, geared mainly to the degradation of patients. Doubtless this happens, and on a vast scale. But they may also be "asylums" in the best sense of the word, a sense perhaps scarcely allowed by Goffman: places that provide a refuge for the tormented, storm-tossed soul, provide it with just that mixture of order and freedom of which it stands in such need. José had suffered from confusion and chaos—partly organic epilepsy, partly the disorder of his life—and from confinement and bondage, also both epileptic and existential. Hospital was good for José, perhaps lifesaving at this point in his life, and there is no doubt that he himself felt this fully.

Suddenly too, after the moral closeness, the febrile intimacy of his house, he now found others, found a world both "professional" and concerned: unjudging, unmoralistic, unaccusing, detached, but at the same time with a real feeling both for him and for his problems. At this point, therefore (he had now been in hospital for four weeks), he started to have hope, started to become more animated, to turn to others as he had never done before—not, at least, since the onset of autism, when he was eight.

But hope, turning to others, interaction, was "forbidden," and no doubt frighteningly complex and "dangerous" as well. José had lived for fifteen years in a guarded, closed world—in what Bruno Bettelheim in his book on autism called the "empty fortress." But it was not, it had never been, for him, entirely empty; there had always been his love for nature, for animals and plants. *This* part of him, *this* door, had always remained open. But now there was temptation, and pressure, to "interact," pressure that was often too much, came too soon. And precisely at such times José would "relapse," would turn again, as if for comfort and security, to the isolation, to the primitive rocking movements, he had at first shown.

The third time I saw José, I did not send for him in the clinic, but went up, without warning, to the admission ward. He was sitting, rocking, in the frightful day room, his face and eyes closed, a picture of regression. I had a qualm of horror when I saw him like this, for I had imagined, had indulged, the notion of "a steady recovery." I had to see José in a regressed condition (as I was to do again and again) to see that there was no simple "awakening" for him, but a path fraught with a sense of danger, double jeopardy, terrifying as well as exciting—because he had come to love his prison bars.

As soon as I called him, he jumped up, and eagerly, hungrily, followed me to the art room. Once more I took a fine pen from my pocket, for he seemed to have an aversion to crayons, which was all they used on the ward. "That fish you drew," I hinted it with a gesture in the air, not knowing how much of my words he might understand, "that fish, can you remember it, can you draw it again?" He nodded eagerly and took the pen from my hands. It was three weeks since he had seen it. What would he draw now?

He closed his eyes for a moment—summoning an image?—

and then drew. It was still a trout, rainbow-spotted, with fringy fins and a forked tail, but, this time, with egregiously human features, an odd nostril (what fish has nostrils?), and a pair of ripely human lips. I was about to take the pen, but, no, he was not finished. What had he in mind? The image was complete. The image, perhaps, but not the scene. The fish before had existed—as an icon—in isolation: now it was to become part of a world, a scene. Rapidly he sketched in a little fish, a companion, swooping into the water, gamboling, obviously in play. And then the surface of the water was sketched in, rising to a sudden, tumultuous wave. As he drew the wave, he became excited, and emitted a strange, mysterious cry.

I couldn't avoid the feeling, perhaps a facile one, that this drawing was symbolic—the little fish and the big fish, perhaps him and me? But what was so important and exciting was the spontaneous representation, the impulse, not my suggestion, entirely from himself, to introduce this new element—a living interplay in what he drew. In his drawings as in his life hitherto, interaction had always been absent. Now, if only in

play, in symbol, it was allowed back. Or was it? What was that angry, avenging wave?

Best to go back to safe ground, I felt; no more free association. I had seen potential, but I had seen, and heard, danger too. Back to safe, Edenic, prelapsarian Mother Nature. I found a Christmas card lying on the table, a robin redbreast on a tree trunk, snow and stark twigs all around. I gestured to the bird and gave José the pen. The bird was finely drawn, and he used a red pen for the breast. The feet were somewhat taloned, grasping the bark (I was struck, here and later, by his need to emphasize the grasping power of hands and feet, to make contact sure, almost gripping, obsessed). But—what was happening?—the dry winter twiglet, next to the tree trunk, had shot up in his drawing, expanded into florid open bloom. There were other things that were perhaps symbolic, although I could not be sure. But the salient and exciting and most significant transformation was this: that José had changed winter into spring.

Now, finally, he started to speak—though "speak" is much too strong a term for the strange-sounding, stumbling, largely unintelligible utterances that came out, on occasion startling him as much as they startled us—for all of us, José included, had regarded him as wholly and incorrigibly mute, whether from incapacity, indisposition, or both (there had been the *attitude*, as well as the fact, of not speaking). And here, too, we found it impossible to say how much was "organic," how much was a matter of "motivation." We had reduced, though not annulled, his temporal lobe disorders—his electroencephalograms (EEGs) were never normal; they still showed in these lobes a sort of low-grade electrical muttering, occasional spikes, dysrhythmia, slow waves. But they were immensely

improved compared with what they were when he came in. If he could remove their convulsiveness, he could not reverse the damage they had sustained.

We had improved, it could not be doubted, his physiological *potentials* for speech, though there was an impairment of his abilities to use, understand, and recognize speech, with which, doubtless, he would always have to contend. But, equally important, he now was fighting for the recovery of his understanding and speech (egged on by all of us, and guided by the speech therapist in particular), where previously he had accepted it, hopelessly or masochistically, and indeed had turned against virtually all communication with others, verbal and otherwise. Speech impairment and the refusal to speak had coupled before in the double malignancy of disease;

now, recovery of speech and attempts to speak were being happily coupled in the double benignity of beginning to get well. Even to the most sanguine of us it was very apparent that José would never speak with any facility approaching normal, that speech could never, for him, be a real vehicle for self-expression, could serve only to express his simpler needs. And he himself seemed to feel this too and, while he continued to fight for speech, turned more fiercely to drawing for self-expression.

One final episode. José had been moved off the frenzied admission ward to a calmer, quieter special ward, more homelike, less prisonlike, than the rest of the hospital: a ward with an exceptional number and quality of staff, designed especially, as Bettelheim would say, as "a home for the heart," for patients with autism who seem to require a kind of loving and dedicated attention that few hospitals can give. When I went up to this new ward, he waved his hand lustily as soon as he saw me—an outgoing, open gesture. I could not imagine him having done this before. He pointed to the locked door, he wanted it open, he wanted to go outside.

He led the way downstairs, outside, into the overgrown, sunlit garden. So far as I could learn, he had not voluntarily gone outside since he was eight, since the very start of his illness and withdrawal. Nor did I have to offer him a pen—he took one himself. We walked around the hospital grounds, José sometimes gazing at the sky and trees, but more often down at his feet, at the mauve and yellow carpet of clover and dandelions beneath us. He had a very quick eye for plant forms and colors, rapidly saw and picked a rare white clover, and found a still rarer four-leaf one. He found seven different types of grass, no less, seemed to recognize, to greet, each one as a friend. He was delighted most of all by the great yellow dandelions, open, all their florets flung open to the sun. This was his plant—it was how he felt, and to show his feeling he would draw it. The need to draw, to pay graphic reverence, was immediate and strong: he knelt down, placed his clip-board on the ground, and, holding the dandelion, drew it.

This, I think, is the first drawing from real life that José had done since his father took him sketching as a child, before he became ill. It is a splendid drawing, accurate and alive. It shows his love for reality, for another form of life. It is, to my mind, rather similar to, and not inferior to, the fine vivid flowers one finds in medieval botanies and herbals—fastidiously, botanically exact, even though José has no formal knowledge of botany, and could not be taught it or understand it if he tried. His mind is not built for the abstract, the conceptual. *That* is not available to him as a path to truth. But he has a

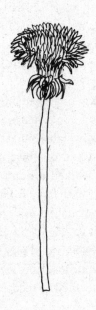

passion and a real power for the particular—he loves it, he enters into it, he re-creates it. And the particular, if one is particular enough, is also a road—one might say nature's road—to reality and truth.

The abstract, the categorical, has no interest for the autistic person—the concrete, the particular, the singular, is all. Whether this is a question of capacity or disposition, it is strikingly the case. Lacking, or indisposed to, the general, the autistic seem to compose their world picture entirely of particulars. Thus they live, not in a universe, but in what William James called a "multiverse," of innumerable, exact, and passionately intense particulars. It is a mode of mind at the opposite extreme from the generalizing, the scientific, but still "real," equally real, in a quite different way. Such a mind has been imagined in Borges's story "Funes the Memorious" (so like Luria's *Mnemonist*):

He was, let us not forget, almost incapable of ideas of a general, Platonic sort. . . . In the teeming world of Funes, there were only details, almost immediate in their presence. . . . No one . . . has felt the heat and pressure of a reality as indefatigable as that which day and night converged upon the hapless Ireneo.

As for Borges's Ireneo, so for José. But it is not necessarily a hapless circumstance: there may be a deep satisfaction in particulars, especially if they shine, as they may do for José, with an emblematic radiance.

I think José, an autist, a simpleton too, has such a gift for the concrete, for *form*, that he is, in his way, a naturalist and natural artist. He grasps the world as forms—directly and intensely felt forms—and reproduces them. He has fine literal powers, but he has figurative powers too. He can draw a flower or fish with remarkable accuracy, but he can also make one which is a personification, an emblem, a dream, or a joke. And the autistic are supposed to lack imagination, playfulness, art!

Creatures like José are not supposed to exist. Autistic child-artists like Lorna Selfe's Nadia were not supposed to exist. Are they indeed so rare, or are they overlooked? Nigel Dennis, in a brilliant essay on *Nadia* in the *New York Review of Books*, wonders how many of the world's "Nadias" may be dismissed or overlooked, their remarkable productions crumpled up and consigned to the trash can, or simply, like José, treated without thought, as an odd talent, isolated, irrelevant, of no interest. But the autistic artist or (to be less lofty) the autistic imagination is by no means rare. I have seen a dozen examples of it in as many years, and this without making any particular effort to find them.

The autistic, by their nature, are seldom open to influence. It is their "fate" to be isolated, and thus original. Their "vision," if it can be glimpsed, comes from within and appears aboriginal. They seem to me, as I see more of them, to be a strange species in our midst, odd, original, wholly inwardly directed, unlike others.

Autism was once seen as a childhood schizophrenia, but phenomenologically the reverse is the case. The schizophrenic's complaint is always of "influence" from the outside: he is passive, he is played upon, he cannot be himself. The autistic would complain—if they complained—of absence of influence, of absolute isolation.

"No man is an island, entire of itself," wrote Donne. But this is precisely what autism is—an island, cut off from the main. In "classical" autism, which is manifest and often total by the third year of life, the cutting off is so early there may be no memory of the main. In "secondary" autism, like José's, caused by brain disease at a later stage in life, there is some memory, perhaps some nostalgia, for the main. This may explain why José was more accessible than most, and why, at least in drawing, he may show interplay taking place.

Is being an island, being cut off, necessarily a death? It may be a death, but it is not necessarily so. For though "horizontal" connections with others, with society and culture, are lost, yet there may be vital and intensified "vertical" connections, direct connections with nature, with reality, uninfluenced, unmediated, untouchable, by any others. This "vertical" contact is very striking with José, hence the piercing directness, the absolute clarity of his perceptions and drawings, without a hint or shade of ambiguity or indirection, a rocklike power uninfluenced by others.

This brings us to our final question: is there any "place" in the world for a man who is like an island, who cannot be acculturated, made part of the main? Can "the main" accommodate, make room for, the singular? There are similarities here to the social and cultural reactions to genius. (Of course, I do not suggest that all autists have genius, only that they share with genius the problem of singularity.) Specifically: what does the future hold for José? Is there some "place" for him in the world which will *employ* his autonomy, but leave it intact?

Could he, with his fine eye, and great love of plants, make illustrations for botanical works or herbals? Be an illustrator for zoology or anatomy texts? (See the drawing, on the next page, he made for me when I showed him a textbook illustration of the layered tissue called "ciliated epithelium.") Could he accompany scientific expeditions and make drawings (he paints and makes models with equal facility) of rare species? His pure concentration on the thing before him would make him ideal in such situations.

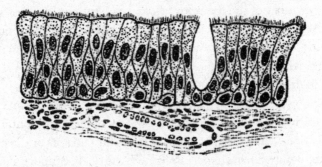

Ciliated epithelium from the trachea of a kitten (magnified 255 times).

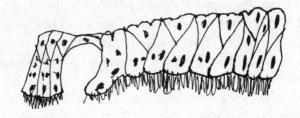

Or, to take a strange but not illogical leap, could he, with his peculiarities, his idiosyncrasy, do drawings for fairy tales, nursery tales, Bible tales, myths? Or (since he cannot read, and sees letters only as pure and beautiful forms) could he not illustrate, and elaborate, the gorgeous capitals of manuscript breviaries and missals? He has done beautiful altarpieces, in mosaic and stained wood, for churches. He has carved exquisite lettering on tombstones. His current "job" is hand printing sundry notices for the ward, which he does with the flourishes and elaborations of a latter-day Magna Carta. All this he could do, and do very well. And it would be of use and delight to others, and delight him too. He could do all of these—but, alas, he will do none, unless someone very under-standing, and with opportunities and means, can guide and employ him. For, as the stars stand, he will probably do nothing, and spend a useless, fruitless life, as so many other autistic people do, overlooked, unconsidered, in the back ward of a state hospital.

POSTSCRIPT

After publication of this piece, I again received many offprints and letters, the most interesting being from Dr. Clara Claiborne Park. It is indeed clear (as Nigel Dennis suspected) that even though Nadia may have been unique, a sort of Picasso, artistic gifts of fairly high order are not uncommon among the autistic. Testing for artistic potential, as in the Goodenough "Draw-a-Man" intelligence test, is almost useless: there must occur, as with Nadia, José, and the Parks' daughter Ella, a *spontaneous* production of striking drawings.

In an important and richly illustrated review of *Nadia*, Dr. Park brings out, on the basis of experience with her own child no less than from a perusal of the world literature, what seem to be the cardinal characteristics of such drawings. These include "negative" characteristics, such as derivativeness and stereotypy, and "positive" ones, such as an unusual capacity for delayed rendition, and for rendering the object as *perceived* (not as *con*ceived): hence the sort of inspired naïveté especially seen. She also notes a relative indifference to display of others' reactions, which might seem to render such children untrainable. And yet, manifestly, this need not be the case. Such children are not necessarily unresponsive to teaching or attention, though this may need to be of a very special type.

In addition to experience with her own child, who is now an accomplished adult artist, Dr. Park cites also the fascinating and insufficiently known experiences of the Japanese, especially Morishima and Motzugi, who have had remarkable success in bringing autists from an untutored (and seemingly unteachable) childhood giftedness to professionally accomplished adult artistry. Morishima favors special instructional techniques ("highly structured skill training"), a sort of apprenticeship in the classical Japanese cultural tradition, and encouragement of drawing *as a means of communication*. But such formal training, though crucial, is not enough. A most intimate, empathic relationship is required. The words with which Dr. Park concludes her review may properly conclude "The World of the Simple":

> The secret may lie elsewhere, in the dedication that led Motzugi to live with another retarded artist in his home, and to write: "The secret in developing Yanamura's talent was to share his spirit. The teacher should love the beautiful, honest retarded person, and live with a purified, retarded world."

Annotated Bibliography

General References

Hughlings Jackson, Kurt Goldstein, Henry Head, A. R. Luria—
these are the fathers of neurology who lived intensely, and thought
intensely, about patients and problems not so dissimilar to our
own. They are always present in the neurologist's mind, and they
haunt the pages of this book. There is a tendency to reduce complex figures to stereotypes, to disallow the fullness, and often the
rich contradictoriness, of their thought. Thus, I often talk about
classical "Jacksonian" neurology, but the Hughlings Jackson who
wrote of "dreamy states" and "reminiscence" was very different
from the Jackson who saw all thought as propositional calculus.
The former was a poet, the latter a logician, and yet they are one
and the same man. Henry Head the diagram-maker, with his passion for schematics, was very different from the Head who wrote
poignantly of "feeling-tone." Goldstein, who wrote so abstractly
of "the Abstract," delighted in the rich concreteness of individual
cases. In Luria, finally, the doubleness was conscious; he had,
he felt, to write two sorts of books: formal structural books like
Higher Cortical Functions in Man and biographical "novels" like
The Mind of a Mnemonist. The first he called "Classical Science,"
the second, "Romantic Science."

Jackson, Goldstein, Head, and Luria—they constitute the
essential axis of neurology, and certainly they are the axis of my
own thinking and of this book. My first references must therefore
be to them—ideally to everything they wrote, for what is most
characteristic is always suffused through a life's work—but for

the sake of practicality to certain key works which are the most accessible to English-speaking readers.

Hughlings Jackson

There are wonderful descriptions of cases before Hughlings Jackson—such as Parkinson's "Essay on the Shaking Palsy," as early as 1817—but no general vision or systemization of nervous function. Jackson is the founder of neurology as a science. One can browse through the basic volumes of Jacksoniana: *Selected Writings of John Hughlings Jackson*, edited by James Taylor, Gordon Holmes, and F. M. R. Walshe (London: Hodder and Stoughton, 1932). These writings are not easy reading, though often evocative and dazzlingly clear in parts. A further selection, with records of Jackson's conversations and a memoir, had been almost completed by Purdon Martin at the time of his death, and will, it is hoped, be published in this sesquicentennial year of Jackson's birth.

Henry Head

Head, like Weir Mitchell (see below under Chapter 6), is a marvelous writer, and *his* heavy volumes, unlike Jackson's, are always a delight to read:

Studies in Neurology. 2 vols. Oxford: Oxford University Press, 1920.
Aphasia and Kindred Disorders of Speech. 2 vols. Cambridge: Cambridge University Press, 1926.

Kurt Goldstein

Goldstein's most accessible general book is *The Organism: A Holistic Approach to Biology Derived from Pathological Data in Man* (New York: American Book Company, 1939). See also K. Goldstein and M. Sheerer, "Abstract & concrete behaviour," *Psychological Monographs* 53 (1941).

Goldstein's fascinating case histories, scattered through many books and journals, await collection.

A. R. Luria

The greatest neurological treasures of our time, for both thought and case description, are the works of A. R. Luria. Most of Luria's books have been translated into English. The most accessible are

The Man with a Shattered World. New York: Basic Books, 1972.
The Mind of a Mnemonist. New York: Basic Books, 1968.
Speech and the Development of Mental Processes in the Child. Coauthored with F. Ia. Yudovich. London: Staples Press, 1959. A study of mental defect, speech, play, and twins.
Human Brain and Psychological Process. New York: Harper & Row, 1966. Case histories of patients with frontal lobe syndromes.
The Neuropsychology of Memory. New York: John Wiley & Sons, 1976.
Higher Cortical Functions in Man. 2nd ed. New York: 1980. Luria's magnum opus—the greatest synthesis of neurological work and thought in our century.
The Working Brain: An Introduction of Neuropsychology. New York: Basic Books, 1973. A condensed and highly readable version of the above. The best available introduction to neuropsychology.

Chapter References and Suggested Reading

1. The Man Who Mistook His Wife for a Hat

Damasio, Antonio R. "Disorders in Visual Processing," in M. M. Mesulam, *Principles of Behavioral Neurology* (1985), pp. 259–88.
Kertesz, Andrew. "Visual agnosia: the dual deficit of perception and recognition." *Cortex* (1979) 15: 403–19.
Macrae, Donald, and Elli Trolle. "The defect of function in visual agnosia." *Brain* (1956) 79 (1): 94–110.
Marr, David. *Vision: A Computational Investigation of Visual Representation in Man*. San Francisco: W. H. Freeman, 1982. (See further description below under Chapter 15.)

2. The Lost Mariner

Korsakov's original (1887) contribution and his later works have not been translated. A full bibliography, with translated excerpts and discussion, is given in A. R. Luria's *Neuropsychology of Memory*, which itself provides many striking examples of amnesia akin to that of "The Lost Mariner." Both here, and in the preceding case history, I refer to Anton, Pötzl, and Freud. Of these only Freud's monograph—a work of great importance—has been translated into English.

Anton, Gabriel. "Über die Selbstwarnehmung der Herderkrankungen des Gehirns durch den Kranken." *Arch. Psychiat.* (1899) 32.

Freud, Sigmund. *Zur Auffassung der Aphasia*. Leipzig: 1891. Translated by E. Stengel as *On Aphasia: A Critical Study*. New York: International Universities Press, 1953.

Pötzl, Otto. *Die Aphasielehre vom Standpunkt der klinishcen Psychiatrie: Die Optische-agnostischen Störungen*. Leipzig: 1928. The syndrome Pötzl describes is not merely visual, but may extend to a complete unawareness of parts, or one half, of the body. As such it is also relevant to the themes of Chapters 3, 4, and 8. It is also referred to in my book *A Leg to Stand On* (1984).

3. The Disembodied Lady

Cole, Jonathan. *Pride and a Daily Marathon*. London: Duckworth, 1991.

Martin, James Purdon. *The Basal Ganglia and Posture*. Philadelphia: J. B. Lippincott, 1967.

Mitchell, Silas Weir. *Injuries of Nerves*. 1872; Dover reprint, 1965. (See further description below under Chapter 6.)

Sherrington, Charles S. *The Integrative Action of the Nervous System*. Cambridge: Cambridge University Press, 1906. See especially pages 335–43.

———. *Man on His Nature*. Cambridge: Cambridge University Press, 1940. Chapter 11, especially pages 328–29, has the most direct relevance to this patient's condition.

Sterman, Arnold B., Herbert H. Schaumburg, and Arthur K. Asbury. "The acute sensory neuronopathy syndrome." *Annals of Neurology* (1980) 7: 354–58.

4. The Man Who Fell Out of Bed

Pötzl, Otto. *Die Aphasielehre vom Standpunkt der klinishcen Psychiatrie: Die Optische-agnostischen Störungen.* Leipzig: 1928. (Also see above in Chapter 2.)

5. Hands

Leont'ev, A. N., and A. V. Zaporozhets. *Rehabilitation of Hand Function.* New York: Pergamon Press, 1960.

6. Phantoms

Cole, Jonathan, H. A. Katifi, and E. M. Sedgwick. "Observations on a man without large myelinated sensory fibre input from below the neck." *Journal of Physiology* (1986). 376, 47P.

Mitchell, Silas Weir. *Injuries of Nerves.* 1872; Dover reprint, 1965. This great book contains Weir Mitchell's classic accounts of phantom limbs, reflex paralysis, etc., from the American Civil War. It is wonderfully vivid and easy to read, for Weir Mitchell was a novelist no less than a neurologist. Indeed, some of his most imaginative neurological writings (such as "The Case of George Dedlow") were published not in scientific journals but in the *Atlantic Monthly* in the 1860s and 1870s, and are therefore not very accessible now, though they enjoyed an immense readership at the time.

Sterman, Arnold B., et al. See above under Chapter 3.

7. On the Level

Martin, James Purdon. *The Basal Ganglia and Posture.* Philadelphia: J. B. Lippincott, 1967.

8. Eyes Right!

Battersby, William S., Morris B. Bender, Max Pollack, and Robert L. Kahn. "Unilateral 'spatial agnosia' ('inattention') in patients with cerebral lesions." *Brain* (1956) 79 (1): 68–93.

Mesulam, M. Marsel. *Principles of Behavioral Neurology* (Philadelphia: F. A. Davis, 1985), pp. 259–88.

9. The President's Speech

The best discussion of Frege on "tone" is to be found in Michael Dummett's *Frege: Philosophy of Language* (London: Duckworth, 1973), especially pages 83–89.

Henry Head's discussion of speech and language, in particular its "feeling-tone," is best read in his treatise on aphasia.

Hughlings Jackson's work on speech was widely scattered, but much was brought together posthumously in "Hughlings Jackson on aphasia and kindred affections of speech, together with a complete bibliography of his publications of speech and a reprint of some of the more important papers," *Brain* (1915) 38: 1–190.

On the complex and confused subject of the auditory agnosias, see Henri Hécaen and Martin L. Albert, *Human Neuropsychology* (New York: Wiley, 1978), pp. 265–76.

10. Witty Ticcy Ray

In 1885 Gilles de la Tourette published a two-part paper in which he described with extreme vividness (he was a playwright as well as a neurologist) the syndrome that now bears his name: "Etude sur an affection nerveuse caracterisée par l'incoordination mortice accompagnée d'echolalie et de coprolalie," *Archives de Neurologie* 9: 19–42, 158–200. The first English translation of these papers, with interesting editorial comments, is provided by: C. G. Goetz and H. L. Klawans, "Gilles de la Tourette on Tourette Syndrome," *Advances in Neurology* (1982) 35: 1–16.

In Henri Meige and Eugene Feidel's great *Les Tics et leur traitement* (1902), brilliantly translated by Kinnier Wilson in 1907 (London: Appleton), there is a wonderful opening personal memoir by a patient, "Les confidences d'un ticqueur," which is unique of its kind.

11. Cupid's Disease

As with Tourette's syndrome, we must go back to the older literature to find full clinical descriptions. Emil Kraepelin, Freud's contemporary, provides many striking vignettes of neurosyphilis. The interested reader might consult Kraepelin's *Lectures on Clinical Psychiatry* (New York: William Wood, 1904), in particular Chapters 10 and 12, on megalomania and delirium in general paralysis.

12. A Matter of Identity

Luria, A. R. *The Neuropsychology of Memory*. New York: John Wiley & Sons, 1976.

13. Yes, Father-Sister

Luria, A. R. *Human Brain and Psychological Process*. New York: Harper & Row, 1966.

14. The Possessed

See references above under Chapter 10, "Witty Ticcy Ray."

15. Reminiscence

Alajouanine, Théophile. "Dostoievski's epilepsy." *Brain* (1963) 86: 209–21.

Critchley, Macdonald, and R. A. Henson, eds. *Music and the Brain: Studies in the Neurology of Music*. London: Charles C. Thomas, 1977. See especially chapters 19 and 20.

Marr, David. *Vision: A Computational Investigation of Visual Representation in Man*. San Francisco: W. H. Freeman, 1982. This is a work of extreme originality and importance, published posthumously (Marr contracted leukemia while still a young man). Penfield shows us the forms of the brain's final representations—voices, faces, tunes, scenes—the "iconic": Marr shows us what is not intuitively obvious, or ever normally experienced—the form of the brain's initial representations. Perhaps I should have given this reference in Chapter 1—it is certain that Dr. P. had some "Marr-like" deficits, difficulties in forming what Marr calls a "primal sketch" in addition to, or underlying, his physiognomonic difficulties. Probably no neurological study of imagery, or memory, can dispense with the considerations raised by Marr.

Penfield, Wilder, and Phanor Perot. "The brain's record of visual and auditory experience: a final summary and discussion." *Brain* (1963) 86: 595–696. I regard this magnificent 100-page paper, the culmination of nearly thirty years' profound observation, experiment and thought, as one of the most original and important in all neurology. It stunned me when it came out in 1963 and was constantly in my mind when I wrote *Migraine* in 1967. It is the essential reference and inspiration to the whole of this section. More readable than many novels, it has a wealth and strangeness of material which any novelist would envy.

Salaman, Esther. *A Collection of Moments: A Study of Involuntary Memories*. London: Longman, 1970.

Williams, Denis. "The structure of emotions reflected in epileptic experiences." *Brain* (1956) 79 (1): 29–67.

Hughlings Jackson was the first to address himself to "psychical seizures," to describe their almost novelistic phenomenology and to identify their anatomical loci in the brain. He wrote several papers on the subject. Most pertinent are those published in Volume 1 of his *Selected Writings* (1931), pp. 251ff. and 274ff., and the following (not included in that volume):

Jackson, John Hughlings. "On right- or left-sided spasm at the onset of epileptic paroxysms, and on crude sensation warnings, and elaborate mental states." *Brain* (1880) 3: 192–206.

———. "On a particular variety of epilepsy ('Intellectual Aura')." *Brain* (1888) 11: 179–207.

James Purdon Martin has provided an intriguing suggestion that Henry James met Hughlings Jackson, discussed such seizures with him, and employed this knowledge in his depiction of the uncanny apparitions in *The Turn of the Screw*: "Neurology in fiction: *The Turn of the Screw*," *British Medical Journal* (1973) 4: 717–21.

16. Incontinent Nostalgia

Jelliffe, Smith Ely. *Psychopathology of Forced Movements and Oculogyric Crises of Lethargic Encephalitis*. New York and Washington: Nervous and Mental Disease Publishing Co., 1932. See especially page 114ff. discussing Zutt's paper of 1930.

See also the case of "Rose R." in my own book *Awakenings* (1973).

17. A Passage to India

I am not acquainted with the literature on this subject. I have, however, had personal experience of another patient—also with a glioma, with increased intracranial pressure and seizures, and on steroids—who, as she was dying, had similar nostalgic visions and reminiscences, in her case of the Midwest.

18. The Dog Beneath the Skin

Bear, David. "Temporal-lobe epilepsy: a syndrome of sensory-limbic hyperconnection." *Cortex* (1979) 15: 357–84.

Brill, A. A. "The sense of smell in neuroses and psychoses." *Psychoanalytical Quarterly* (1932) 1: 7–42. Brill's lengthy paper covers much more ground than its title would indicate. In particular it contains a detailed consideration of the strength and the importance of smell in many animals, in "savages," and in children, the amazing powers and potentials of which seem to have been lost in adult man.

19. Murder

I am not acquainted with any precisely similar accounts. I have, however, in rare cases of frontal lobe injury, frontal lobe tumor, frontal lobe (anterior cerebral) "stroke" and (not least) lobotomy, seen the precipitation of obsessional "reminiscence." Lobotomies, of course, were designed as a "cure" for such "reminiscence"—but, on occasion, caused it to become very much worse. See also Penfield and Perot, listed above under Chapter 15.

20. The Visions of Hildegard

Singer, Charles. "The visions of Hildegard of Bingen" in *From Magic to Science: Essays on the Scientific Twilight*. London: Ernest Benn, 1928.

See also my *Migraine* (1970), especially Chapter 3, on migraine aura.

For Dostoyevsky's epileptic transports and visions, see Alajouanine, Théophile. "Dostoievski's epilepsy." *Brain* (1963) 86: 209–21.

Introduction to Part Four

Bruner, Jerome. "Narrative and paradigmatic modes of thought," presented at the Annual Meeting of the American Psychological Association, Toronto, August 1984. Published as "Two Modes of Thought," in *Actual Minds, Possible Worlds* (Boston: Harvard University Press, 1986), pp. 11–43.

Luria, A. R., and F. Ia. Yudovich. *Speech and the Development of Mental Processes in the Child*. London: Staples Press, 1959.

Miller, L. K. "Developmentally delayed musical savant's sensitivity to tonal structure." *American Journal of Mental Deficiency* (1987) 91 (5), 467–71.

Scholem, Gershom. *On the Kabbalah and Its Symbolism*. New York: Schocken Books, 1965.

Yates, Frances. *The Art of Memory*. London: Routledge and Kegan Paul, 1966.

21. Rebecca

Bruner, Jerome. See above under Introduction to Part Four.

Peters, Larry G. "The role of dreams in the life of a mentally retarded individual." *Ethos* (1983) 11 (1–2): 49–65.

22. A Walking Grove

Hill, Lewis. "Idiots savants: a categorization of abilities." *Mental Retardation*. December 1974.

Viscott, David. "A musical idiot savant: a psychodynamic study, and some speculation on the creative process." *Psychiatry* (1970) 33 (4): 494–515.

23. The Twins

Hamblin, D. J. "They are 'idiot savants'—wizards of the calendar." *Life* 60 (18 March 1966): 106–8.

Horwitz, William A., Clarice Kestenbaum, Ethel Person, and Lissy Jarvik. "Identical twin 'idiots savants'—calendar calculators." *American Journal of Psychiatry* (1965) 121: 1075–79.

Luria, A. R., and F. Ia. Yudovich. *Speech and the Development of Mental Processes in the Child.* Eng. tr. London: 1959.

Myers, F. W. H. *Human Personality and Its Survival of Bodily Death.* London: 1903. See chapter 3, "Genius," especially pages 70–87. Myers was in part a genius, and this book is in part a masterpiece. This is evident in the first volume, which is often comparable to Williams James's *Principle of Psychology*—he was a close personal friend of James. (The second volume, "Phantasms of the Dead," etc., is to my mind an embarrassment.)

Nagel, Ernest, and James R. Newmann. *Gödel's Proof.* New York: New York University Press, 1958.

Park, Clara Claiborne, and David Park. See below under Chapter 24.

Selfe, Lorna. *Nadia.* See below under Chapter 24.

Silverberg, Robert. *Thorns.* New York: Ballantine Books, 1967.

Smith, Steven B. *The Great Mental Calculators: The Psychology, Methods, and Lives of Calculating Prodigies, Past and Present.* New York: Columbia University Press, 1983.

Stewart, Ian. *Concepts of Modern Mathematics.* Harmondsworth: Penguin Books, 1975.

Wollheim, Richard. *The Thread of Life.* Cambridge, Mass.: Harvard University Press, 1984. See especially chapter 3 on "iconicity" and "centricity." I had just read this book when I came to write of Martin A., the Twins, and José; hence, reference to it appears in all three of these chapters (22, 23, 24).

24. The Autist Artist

Buck, Lucien A., Elayne Kardeman, and Fran Goldstein. "Artistic talent in 'autistic' adolescents and young adults." *Empirical Studies of the Arts* (1985) 3 (1): 81–104.

———. "Art as a means of interpersonal communication in autistic young adults." *Journal of Psychology and Christianity* (1984) 3 (3): 73–84. (Both these papers are published under the aegis of the Talented Handicapped Artist's Workshop, founded in New York in 1981.)

Morishima, Akira. "Another Van Gogh of Japan: The superior art work of a retarded boy." *Exceptional Children* (1974) 41: 92–96.

Motsugi, K. "Shyochan's drawing of insects." *Japanese Journal of Mentally Retarded Children* (1968) 119: 44–47.

Park, Clara Claiborne. *The Siege: The First Eight Years of an Autistic Child*. Boston: Little, Brown, 1967. (See also Park's later books and essays on her daughter, Jessy Park.)

Park, David, and P. Youderian. "Light and number: ordering principles in the world of an autistic child." *Journal of Autism and Childhood Schizophrenia* (1974) 4 (4): 313–23.

Rapin, Isabelle. *Children with Brain Dysfunction: Neurology, Cognition, Language and Behaviour*. New York: Raven Press, 1982.

Selfe, Lorna. *Nadia: A Case of Extraordinary Drawing Ability in an Autistic Child*. London: Academic Press, 1977. This richly illustrated study of a uniquely gifted child attracted much attention when published and some very important critiques and reviews. The reader is also referred to Nigel Dennis, *New York Review of Books*, May 4, 1978; and Clara Claiborne Park, *Journal of Autism and Childhood Schizophrenia* (1978) 8: 457–72. The latter contains a rich discussion and bibliography of the fascinating Japanese work with autistic artists with which my final Postscript concludes.

Index

ON THE MOVE
A Life

When Oliver Sacks was twelve years old, a perceptive school-master wrote: "Sacks will go far, if he does not go too far." It is now abundantly clear that Sacks never stopped going. With unbridled honesty and humor, Sacks writes about the passions that have driven his life—from motorcycles and weight lifting to neurology and poetry. He writes about his love affairs, both romantic and intellectual; his guilt over leaving his family to come to America; his bond with his schizophrenic brother; and the writers and scientists— W. H. Auden, Gerald M. Edelman, Francis Crick—who have influenced his work. *On the Move* is the story of a brilliantly unconventional physician and writer, a man who has illuminated the many ways that the brain makes us human.

Biography

ALSO AVAILABLE

An Anthropologist on Mars
Awakenings
Hallucinations
The Island of the Colorblind
A Leg to Stand On
Migraine
The Mind's Eye
Musicophilia
Oaxaca Journal
Seeing Voices
Uncle Tungsten

VINTAGE BOOKS
Available wherever books are sold.
www.vintagebooks.com